Textbook of Mathematics

Textbook of Mathematics

Diana Marks

Larsen & Keller
www.larsen-keller.com

Textbook of Mathematics
Diana Marks
ISBN: 978-1-64172-682-5 (Hardback)

Larsen & Keller
Published by Larsen and Keller Education,
5 Penn Plaza,
19th Floor,
New York, NY 10001, USA

Cataloging-in-Publication Data

Textbook of mathematics / Diana Marks.
     p. cm.
Includes bibliographical references and index.
ISBN 978-1-64172-682-5
1. Mathematics. 2. Mathematics--Textbooks. I. Marks, Diana.
QA37.3 .T49 2022
510--dc23

For more information regarding Larsen and Keller Education and its products, please visit the publisher's website www.larsen-keller.com

TABLE OF CONTENTS

Preface   VII

Chapter 1 Introduction   **1**

- Mathematics   1
- The Nature of Mathematics   4
- Branches of Mathematics   8
- Pure Mathematics   10
- Applied Mathematics   12

Chapter 2 Arithmetic   **18**

- Number Theory   24
- Addition   33
- Subtraction   33
- Multiplication   42
- Division   52
- Decimal   58

Chapter 3 Geometry   **64**

- Euclidean Geometry   66
- Analytic Geometry   72
- Non-euclidean Geometry   80
- Differential Geometry   83
- Projective Geometry   87

Chapter 4 Algebra   **92**

- Elementary Algebra   107
- Abstract Algebra   112
- Universal Algebra   116

Chapter 5 Trigonometry   **117**

- Pythagorean Triple   132
- Pythagorean Theorem   152

- Trigonometric Functions   153
- Inverse Trigonometric Functions   169
- Applications of Trigonometry   170

Chapter 6 Probability and Statistics   172

- Theories of Probability and Statistics   185
- Probability Axioms   231

Permissions

Index

PREFACE

The purpose of this book is to help students understand the fundamental concepts of this discipline. It is designed to motivate students to learn and prosper. I am grateful for the support of my colleagues. I would also like to acknowledge the encouragement of my family.

The discipline of mathematics is concerned with the study of topics such as quantity, space, change and structure. It uses patterns for the formulation of new conjectures. Pure mathematics and applied mathematics are two of the major domains of mathematics. Pure mathematics focuses on the study of mathematical concepts such as quantity and structure. Applied mathematics deals with the application of mathematical concepts in different fields such as computer science, engineering, business, science and industry. Some of the other subdisciplines within this field are arithmetic, geometry, algebra and analysis. The concepts included in this book on mathematics are of utmost significance and bound to provide incredible insights to readers. Some of the diverse topics covered herein address the varied branches that fall under this category. Those with an interest in this field would find it helpful.

A foreword for all the chapters is provided below:

Chapter – Introduction

The study of numbers, basic shapes, sequences of numbers, patterns, etc. is called mathematics. It is bifurcated into pure and applied mathematics. Several branches of mathematics include arithmetic, geometry, trigonometry, integration, mensuration, algebra, etc. This is an introductory chapter which will briefly introduce all the significant aspects of mathematics.

Chapter – Arithmetic

Arithmetic is the branch of mathematics which refers to the study of properties and manipulation of numbers. Number theory, addition, subtraction, multiplication, division, decimal, etc. are some the concepts that fall under its domain. All the diverse concepts of arithmetic have been carefully analyzed in this chapter.

Chapter – Geometry

Geometry is a sub-discipline of mathematics which deals with the study of points, lines, surfaces, shapes, size, relative position of figures, etc. Euclidean geometry, analytical geometry, non-Euclidean geometry, projective geometry, etc. are some of the branches of geometry. The topics elaborated in this chapter will help in gaining a better perspective of geometry.

Chapter – Algebra

Algebra is concerned with the study of mathematical symbols and the postulates used to manipulate these symbols. These symbols are used to represent numbers and quantities. It is divided into elementary algebra, abstract algebra, universal algebra, etc. This chapter closely examines these concepts of algebra to provide an extensive understanding of the subject.

Chapter – Trigonometry

The branch of mathematics which deals with the relation of lines and angles in a triangle is referred to as trigonometry. Some of its basic principles are Pythagorean triple, Pythagorean theorem, trigonometric functions, inverse trigonometric functions, etc. All the diverse principles of trigonometry have been carefully analyzed in this chapter.

Chapter – Probability and Statistics

Probability helps to determine how likely an event can occur. The practice of collection, arrangement, presentation and analysis of numerical data is called statistics. This chapter delves into probability theory, statistical theory, decision theory, estimation theory, Bayes' theorem, probability axioms, etc. to provide an easy understanding of the subject.

Diana Marks

Introduction 1

- **Mathematics**
- **The Nature of Mathematics**
- **Branches of Mathematics**
- **Pure Mathematics**
- **Applied Mathematics**

The study of numbers, basic shapes, sequences of numbers, patterns, etc. is called mathematics. It is bifurcated into pure and applied mathematics. Several branches of mathematics include arithmetic, geometry, trigonometry, integration, mensuration, algebra, etc. This is an introductory chapter which will briefly introduce all the significant aspects of mathematics.

Mathematics

Mathematics is the science that deals with the logic of shape, quantity and arrangement. Math is all around us, in everything we do. It is the building block for everything in our daily lives, including mobile devices, architecture (ancient and modern), art, money, engineering, and even sports.

Since the beginning of recorded history, mathematic discovery has been at the forefront of every civilized society, and in use in even the most primitive of cultures. The needs of math arose based on the wants of society. The more complex a society, the more complex the mathematical needs. Primitive tribes needed little more than the ability to count, but also relied on math to calculate the position of the sun and the physics of hunting.

Several civilizations — in China, India, Egypt, Central America and Mesopotamia — contributed to mathematics as we know it today. The Sumerians were the first people to develop a counting system. Mathematicians developed arithmetic, which includes basic operations, multiplication, fractions and square roots. The Sumerians' system passed through the Akkadian Empire to the Babylonians around 300 B.C. Six hundred years later, in America, the Mayans developed elaborate calendar systems and were skilled astronomers. About this time, the concept of zero was developed.

As civilizations developed, mathematicians began to work with geometry, which computes areas and volumes to make angular measurements and has many practical applications. Geometry is used in everything from home construction to fashion and interior design.

Geometry went hand in hand with algebra, invented in the ninth century by a Persian mathematician, Mohammed ibn-Musa al-Khowarizmi. He also developed quick methods for multiplying and diving numbers, which are known as algorithms — a corruption of his name.

Algebra offered civilizations a way to divide inheritances and allocate resources. The study of algebra meant mathematicians were solving linear equations and systems, as well as quadratics, and delving into positive and negative solutions. Mathematicians in ancient times also began to look at number theory. With origins in the construction of shape, number theory looks at figurate numbers, the characterization of numbers, and theorems.

Math and the Greeks

The study of math within early civilizations was the building blocks for the math of the Greeks, who developed the model of abstract mathematics through geometry. Greece, with its incredible architecture and complex system of government, was the model of mathematic achievement until modern times. Greek mathematicians were divided into several schools:

- The Ionian School, founded by Thales, who is often credited for having given the first deductive proofs and developing five basic theorems in plane geometry.

- The Pythagorean School, founded by Pythagoras, who studied proportion, plane and solid geometry, and number theory.

- The Eleatic School, which included Zeno of Elea, famous for his four paradoxes.

- The Sophist School, which is credited for offering higher education in the advanced Greek cities. Sophists provided instruction on public debate using abstract reasoning.

- The Platonic School, founded by Plato, who encouraged research in mathematics in a setting much like a modern university.

- The School of Eudoxus, founded by Eudoxus, who developed the theory of proportion and magnitude and produced many theorems in plane geometry.

- The School of Aristotle, also known as the Lyceum, was founded by Aristotle and followed the Platonic school.

In addition to the Greek mathematicians listed above, a number of Greeks made an indelible mark on the history of mathematics. Archimedes, Apollonius, Diophantus, Pappus, and Euclid all came from this era.

During this time, mathematicians began working with trigonometry. Computational in nature, trigonometry requires the measurement of angles and the computation of trigonometric functions, which include sine, cosine, tangent, and their reciprocals. Trigonometry relies on the synthetic geometry developed by Greek mathematicians like Euclid. For example, Ptolemy's theorem gives rules for the chords of the sum and difference of angles, which correspond to the sum and difference formulas for sines and cosines. In past cultures, trigonometry was applied to astronomy and the computation of angles in the celestial sphere.

After the fall of Rome, the development of mathematics was taken on by the Arabs, then the Europeans. Fibonacci was one of the first European mathematicians, and was famous for his theories on arithmetic, algebra, and geometry. The Renaissance led to advances that included decimal fractions, logarithms, and projective geometry. Number theory was greatly expanded upon, and theories like probability and analytic geometry ushered in a new age of mathematics, with calculus at the forefront.

Development of Calculus

In the 17th century, Isaac Newton and Gottfried Leibniz independently developed the foundations for calculus. Calculus development went through three periods: anticipation, development and rigorization. In the anticipation stage, mathematicians were attempting to use techniques that involved infinite processes to find areas under curves or maximize certain qualities. In the development stage, Newton and Leibniz brought these techniques together through the derivative and integral. Though their methods were not always logically sound, mathematicians in the 18th century took on the rigorization stage, and were able to justify them and create the final stage of calculus. Today, we define the derivative and integral in terms of limits.

In contrast to calculus, which is a type of continuous mathematics, other mathematicians have taken a more theoretical approach. Discrete mathematics is the branch of math that deals with objects that can assume only distinct, separated value. Discrete objects can be characterized by integers, whereas continuous objects require real numbers. Discrete mathematics is the mathematical language of computer science, as it includes the study of algorithms. Fields of discrete mathematics include combinatorics, graph theory, and the theory of computation.

People often wonder what relevance mathematicians serve today. In a modern world, math such as applied mathematics is not only relevant, it's crucial. Applied mathematics is the branches of mathematics that are involved in the study of the physical, biological, or sociological world. The idea of applied math is to create a group of methods that solve problems in science. Modern areas of applied math include mathematical physics, mathematical biology, control theory, aerospace engineering, and math finance. Not only does applied math solve problems, but it also discovers new problems or develops new engineering disciplines. Applied mathematicians require expertise in many areas of math and science, physical intuition, common sense, and collaboration. The common approach in applied math is to build a mathematical model of a phenomenon, solve the model, and develop recommendations for performance improvement.

While not necessarily an opposite to applied mathematics, pure mathematics is driven by abstract problems, rather than real world problems. Much of what's pursued by pure mathematicians can have their roots in concrete physical problems, but a deeper understanding of these phenomena brings about problems and technicalities. These abstract problems and technicalities are what pure mathematics attempts to solve, and these attempts have led to major discoveries for mankind, including the Universal Turing Machine, theorized by Alan Turing in 1937. The Universal Turing Machine, which began as an abstract idea, later laid the groundwork for the development of the modern computer. Pure mathematics is abstract and based in theory, and is thus not constrained by the limitations of the physical world.

According to one pure mathematician, pure mathematicians prove theorems, and applied mathematicians construct theories. Pure and applied are not mutually exclusive, but they are rooted in different areas of math and problem solving. Though the complex math involved in pure and applied mathematics is beyond the understanding of most average Americans, the solutions developed from the processes have affected and improved the lives of all.

The Nature of Mathematics

Mathematics relies on both logic and creativity, and it is pursued both for a variety of practical purposes and for its intrinsic interest. For some people, and not only professional mathematicians, the essence of mathematics lies in its beauty and its intellectual challenge. For others, including many scientists and engineers, the chief value of mathematics is how it applies to their own work. Because mathematics plays such a central role in modern culture, some basic understanding of the nature of mathematics is requisite for scientific literacy. To achieve this, students need to perceive mathematics as part of the scientific endeavor, comprehend the nature of mathematical thinking, and become familiar with key mathematical ideas and skills.

Patterns and Relationships

Mathematics is the science of patterns and relationships. As a theoretical discipline, mathematics explores the possible relationships among abstractions without concern for whether those abstractions have counterparts in the real world. The abstractions can be anything from strings of numbers to geometric figures to sets of equations. In addressing, say, "Does the interval between prime numbers form a pattern?" as a theoretical question, mathematicians are interested only in finding a pattern or proving that there is none, but not in what use such knowledge might have. In deriving, for instance, an expression for the change in the surface area of any regular solid as its volume approaches zero, mathematicians have no interest in any correspondence between geometric solids and physical objects in the real world.

A central line of investigation in theoretical mathematics is identifying in each field of study a small set of basic ideas and rules from which all other interesting ideas and rules in that field can be logically deduced. Mathematicians, like other scientists, are particularly pleased when previously unrelated parts of mathematics are found to be derivable from one another, or from some more general theory. Part of the sense of beauty that many people have perceived in mathematics lies not in finding the greatest elaborateness or complexity but on the contrary, in finding the greatest economy and simplicity of representation and proof. As mathematics has progressed, more and more relationships have been found between parts of it that have been developed separately—for example, between the symbolic representations of algebra and the spatial representations of geometry. These cross-connections enable insights to be developed into the various parts; together, they strengthen belief in the correctness and underlying unity of the whole structure.

Mathematics is also an applied science. Many mathematicians focus their attention on solving problems that originate in the world of experience. They too search for patterns and relationships, and in the process they use techniques that are similar to those used in doing purely theoretical mathematics. The difference is largely one of intent. In contrast to theoretical mathematicians, applied mathematicians, in the examples given above, might study the interval pattern of prime numbers to develop a new system for coding numerical information, rather than as an abstract problem. Or they might tackle the area/volume problem as a step in producing a model for the study of crystal behavior.

The results of theoretical and applied mathematics often influence each other. The discoveries of theoretical mathematicians frequently turn out—sometimes decades later—to have unanticipated practical value. Studies on the mathematical properties of random events, for example, led to knowledge that later made it possible to improve the design of experiments in the social and natural sciences. Conversely, in trying to solve the problem of billing long-distance telephone users fairly, mathematicians made fundamental discoveries about the mathematics of complex networks. Theoretical mathematics, unlike the other sciences, is not constrained by the real world, but in the long run it contributes to a better understanding of that world.

Mathematics, Science and Technology

Because of its abstractness, mathematics is universal in a sense that other fields of human thought are not. It finds useful applications in business, industry, music, historical scholarship, politics, sports, medicine, agriculture, engineering, and the social and natural sciences. The relationship between mathematics and the other fields of basic and applied science is especially strong. This is so for several reasons, including the following:

- The alliance between science and mathematics has a long history, dating back many centuries. Science provides mathematics with interesting problems to investigate, and mathematics provides science with powerful tools to use in analyzing data. Often, abstract patterns that have been studied for their own sake by mathematicians have turned out much later to be very useful in science. Science and mathematics are both trying to discover general patterns and relationships, and in this sense they are part of the same endeavor.

- Mathematics is the chief language of science. The symbolic language of mathematics has turned out to be extremely valuable for expressing scientific ideas unambiguously. The statement that $a=F/m$ is not simply a shorthand way of saying that the acceleration of an object depends on the force applied to it and its mass; rather, it is a precise statement of the quantitative relationship among those variables. More important, mathematics provides the grammar of science—the rules for analyzing scientific ideas and data rigorously.

- Mathematics and science have many features in common. These include a belief in understandable order; an interplay of imagination and rigorous logic; ideals of honesty and openness; the critical importance of peer criticism; the value placed on being the first to make a key discovery; being international in scope; and even, with the development of powerful electronic computers, being able to use technology to open up new fields of investigation.

- Mathematics and technology have also developed a fruitful relationship with each other. The mathematics of connections and logical chains, for example, has contributed greatly to the design of computer hardware and programming techniques. Mathematics also contributes more generally to engineering, as in describing complex systems whose behavior can then be simulated by computer. In those simulations, design features and operating conditions can be varied as a means of finding optimum designs. For its part, computer technology has opened up whole new areas in mathematics, even in the very nature of proof, and it also continues to help solve previously daunting problems.

Mathematical Inquiry

Using mathematics to express ideas or to solve problems involves at least three phases: (1) representing some aspects of things abstractly, (2) manipulating the abstractions by rules of logic to find new relationships between them, and (3) seeing whether the new relationships say something useful about the original things.

Abstraction and Symbolic Representation

Mathematical thinking often begins with the process of abstraction—that is, noticing a similarity between two or more objects or events. Aspects that they have in common, whether concrete or hypothetical, can be represented by symbols such as numbers, letters, other marks, diagrams, geometrical constructions, or even words. Whole numbers are abstractions that represent the size of sets of things and events or the order of things within a set. The circle as a concept is an abstraction derived from human faces, flowers, wheels, or spreading ripples; the letter A may be an abstraction for the surface area of objects of any shape, for the acceleration of all moving objects, or for all objects having some specified property; the symbol + represents a process of addition, whether one is adding apples or oranges, hours, or miles per hour. And abstractions are made not only from concrete objects or processes; they can also be made from other abstractions, such as kinds of numbers (the even numbers, for instance).

Such abstraction enables mathematicians to concentrate on some features of things and relieves them of the need to keep other features continually in mind. As far as mathematics is concerned, it does not matter whether a triangle represents the surface area of a sail or the convergence of two lines of sight on a star; mathematicians can work with either concept in the same way. The resulting economy of effort is very useful—provided that in making an abstraction, care is taken not to ignore features that play a significant role in determining the outcome of the events being studied.

Manipulating Mathematical Statements

After abstractions have been made and symbolic representations of them have been selected, those symbols can be combined and recombined in various ways according to precisely defined rules. Sometimes that is done with a fixed goal in mind; at other times it is done in the context of experiment or play to see what happens. Sometimes an appropriate manipulation can be identified easily from the intuitive meaning of the constituent words and symbols; at other times a useful series of manipulations has to be worked out by trial and error.

Typically, strings of symbols are combined into statements that express ideas or propositions. For example, the symbol A for the area of any square may be used with the symbol s for the length of the square's side to form the proposition $A = s2$. This equation specifies how the area is related to the side—and also implies that it depends on nothing else. The rules of ordinary algebra can then be used to discover that if the length of the sides of a square is doubled, the square's area becomes four times as great. More generally, this knowledge makes it possible to find out what happens to the area of a square no matter how the length of its sides is changed, and conversely, how any change in the area affects the sides.

Mathematical insights into abstract relationships have grown over thousands of years, and they are still being extended—and sometimes revised. Although they began in the concrete experience of counting and measuring, they have come through many layers of abstraction and now depend much more on internal logic than on mechanical demonstration. In a sense, then, the manipulation of abstractions is much like a game: Start with some basic rules, then make any moves that fit those rules—which includes inventing additional rules and finding new connections between old rules. The test for the validity of new ideas is whether they are consistent and whether they relate logically to the other rules.

Application

Mathematical processes can lead to a kind of model of a thing, from which insights can be gained about the thing itself. Any mathematical relationships arrived at by manipulating abstract statements may or may not convey something truthful about the thing being modeled. For example, if 2 cups of water are added to 3 cups of water and the abstract mathematical operation 2+3 = 5 is used to calculate the total, the correct answer is 5 cups of water. However, if 2 cups of sugar are added to 3 cups of hot tea and the same operation is used, 5 is an incorrect answer, for such an addition actually results in only slightly more than 4 cups of very sweet tea. The simple addition of volumes is appropriate to the first situation but not to the second—something that could have been predicted only by knowing something of the physical differences in the two situations. To be able to use and interpret mathematics well, therefore, it is necessary to be concerned with more than the mathematical validity of abstract operations and to also take into account how well they correspond to the properties of the things represented.

Sometimes common sense is enough to enable one to decide whether the results of the mathematics are appropriate. For example, to estimate the height 20 years from now of a girl who is 5' 5" tall and growing at the rate of an inch per year, common sense suggests rejecting the simple "rate times time" answer of 7' 1" as highly unlikely, and turning instead to some other mathematical model, such as curves that approach limiting values. Sometimes, however, it may be difficult to know just how appropriate mathematical results are—for example, when trying to predict stock-market prices or earthquakes.

Often a single round of mathematical reasoning does not produce satisfactory conclusions, and changes are tried in how the representation is made or in the operations themselves. Indeed, jumps are commonly made back and forth between steps, and there are no rules that determine how to proceed. The process typically proceeds in fits and starts, with many wrong turns and dead ends. This process continues until the results are good enough.

But what degree of accuracy is good enough? The answer depends on how the result will be used, on the consequences of error, and on the likely cost of modeling and computing a more accurate answer. For example, an error of 1 percent in calculating the amount of sugar in a cake recipe could be unimportant, whereas a similar degree of error in computing the trajectory for a space probe could be disastrous. The importance of the "good enough" question has led, however, to the development of mathematical processes for estimating how far off results might be and how much computation would be required to obtain the desired degree of accuracy.

Branches of Mathematics

Arithmetic

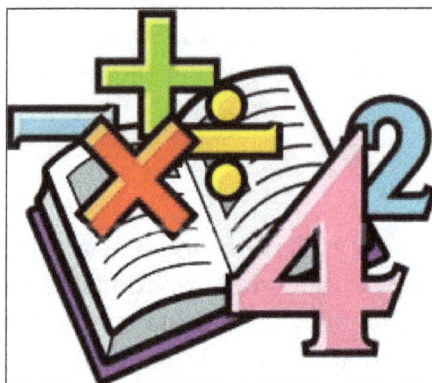

Arithmetic or arithmetics is the oldest and most elementary branch of mathematics, used by almost everyone, for tasks ranging from simple day-to-day counting to advanced science and business calculations. It involves the study of quantity, especially as the result of combining numbers. In common usage, it refers to the simpler properties when using the traditional operations of addition, subtraction, multiplication and division with smaller values of numbers. Professional mathematicians sometimes use the term (higher) arithmetic when referring to more advanced results related to number theory, but this should not be confused with elementary arithmetic.

Geometry

Geometry or "Earth-Measuring" is a part of mathematics concerned with questions of size, shape, relative position of figures, and the properties of space. Geometry is one of the oldest sciences. Initially a body of practical knowledge concerning lengths, areas, and volumes, in the 3rd century BC geometry was put into anaxiomatic form by Euclid, whose treatment—Euclidean geometry—set a standard for many centuries to follow. The field of astronomy, especially mapping the positions of the stars and planets on the celestial sphere, served as an important source of geometric problems during the next one and a half millennia. A mathematician who works in the field of geometry is called a geometer.

Trigonometry

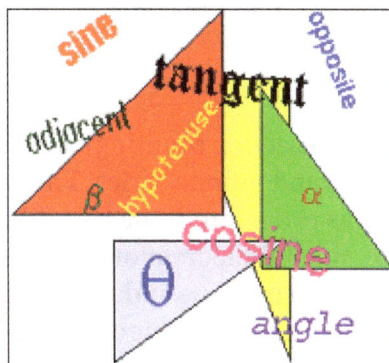

Trigonometry is a branch of mathematics that studies triangles, particularly right triangles. Trigonometry deals with relationships between the sides and the angles of triangles and with the trigonometric functions, which describe those relationships, as well as describing angles in general and the motion of waves such as sound and light waves.

Mensuration

That branch of applied geometry which gives rules for finding the length of lines, the areas of surfaces, or the volumes of solids, from certain simple data of lines and angles.

Algebra

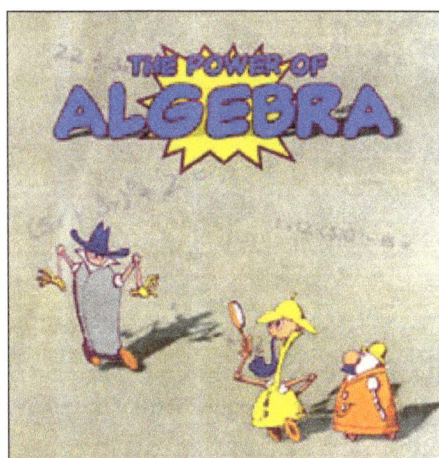

Algebra is the branch of mathematics concerning the study of the rules of operations and relations, and the constructions and concepts arising from them, including terms, polynomials, equations and algebraic structures. Together with geometry, analysis, topology, combinatorics, and number theory, algebra is one of the main branches of pure mathematics. The part of algebra called elementary algebra is often part of the curriculum in secondary education and introduces the concept of variables representing numbers. Statements based on these variables are manipulated using the rules of operations that apply to numbers, such as addition. This can be done for a variety of reasons, including equation solving. Algebra is much broader than elementary algebra and studies what happens when different rules of operations are used and when operations are devised for things other than numbers. Addition and multiplication can be generalized and their precise definitions lead to structures such as groups, rings and fields.

Calculus

Calculus (Latin, calculus, a small stone used for counting) is a branch in mathematics focused on limits, functions, derivatives, integrals, and infinite series. This subject constitutes a major part of modern mathematics education. It has two major branches, differential calculus and integral calculus, which are related by the fundamental theorem of calculus. Calculus is the study of change, in the same way that geometry is the study of shape and algebra is the study of operations and their application to solving equations. A course in calculus is a gateway to other, more advanced courses in mathematics devoted to the study of functions and limits, broadly called mathematical analysis. Calculus has widespread applications in science, economics, and engineering and can solve many problems for which algebra alone is insufficient.

Pure Mathematics

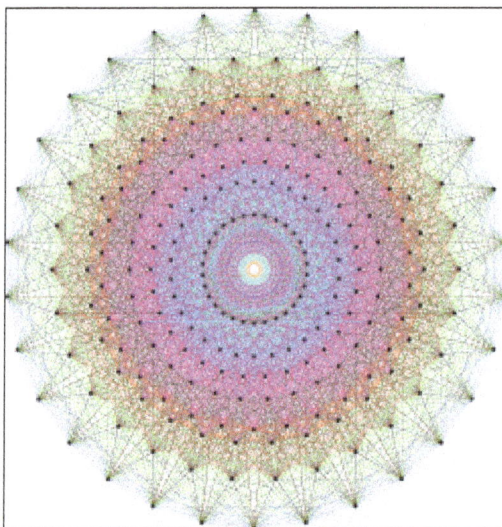

Pure mathematics studies the properties and structure of abstract objects, such as the E8 group, in group theory. This may be done without focusing on concrete applications of the concepts in the physical world.

Pure mathematics is the study of mathematical concepts independently of any application outside mathematics. These concepts may originate in real-world concerns, and the results obtained may later turn out to be useful for practical applications, but the pure mathematicians are not primarily motivated by such applications. Instead, the appeal is attributed to the intellectual challenge and aesthetic beauty of working out the logical consequences of basic principles.

While pure mathematics has existed as an activity since at least Ancient Greece, the concept was elaborated upon around the year 1900, after the introduction of theories with counter-intuitive properties (such as non-Euclidean geometries and Cantor's theory of infinite sets), and the discovery of apparent paradoxes (such as continuous functions that are nowhere differentiable, and Russell's paradox). This introduced the need of renewing the concept of mathematical rigor and rewriting all mathematics accordingly, with a systematic use of axiomatic methods. This led many mathematicians to focus on mathematics for its own sake, that is, pure mathematics.

Nevertheless, almost all mathematical theories remained motivated by problems coming from the real world or from less abstract mathematical theories. Also, many mathematical theories, which had seemed to be totally pure mathematics, were eventually used in applied areas, mainly physics and computer science. A famous early example is Isaac Newton's demonstration that his law of universal gravitation implied that planets move in orbits that are conic sections, geometrical curves that had been studied in antiquity by Apollonius. Another example is the problem of factoring large integers, which is the basis of the RSA cryptosystem, widely used to secure internet communications.

It follows that, presently, the distinction between pure and applied mathematics is more a philosophical point of view or a mathematician's preference than a rigid subdivision of mathematics. In particular, it is not uncommon that some members of a department of applied mathematics describe themselves as pure mathematicians.

Generality and Abstraction

An illustration of the Banach–Tarski paradox, a famous result in pure mathematics. Although it is proven that it is possible to convert one sphere into two using nothing but cuts and rotations, the transformation involves objects that cannot exist in the physical world.

One central concept in pure mathematics is the idea of generality; pure mathematics often exhibits a trend towards increased generality. Uses and advantages of generality include the following:

- Generalizing theorems or mathematical structures can lead to deeper understanding of the original theorems or structures.

- Generality can simplify the presentation of material, resulting in shorter proofs or arguments that are easier to follow.

- One can use generality to avoid duplication of effort, proving a general result instead of having to prove separate cases independently, or using results from other areas of mathematics.

- Generality can facilitate connections between different branches of mathematics. Category theory is one area of mathematics dedicated to exploring this commonality of structure as it plays out in some areas of math.

Generality's impact on intuition is both dependent on the subject and a matter of personal preference or learning style. Often generality is seen as a hindrance to intuition, although it can certainly function as an aid to it, especially when it provides analogies to material for which one already has good intuition.

As a prime example of generality, the Erlangen program involved an expansion of geometry to accommodate non-Euclidean geometries as well as the field of topology, and other forms of geometry, by viewing geometry as the study of a space together with a group of transformations. The study of numbers, called algebra at the beginning undergraduate level, extends to abstract algebra at a more advanced level; and the study of functions, called calculus at the college freshman level becomes mathematical analysis and functional analysis at a more advanced level. Each of these branches of more *abstract* mathematics have many sub-specialties, and there are in fact many connections between pure mathematics and applied mathematics disciplines. A steep rise in abstraction was seen mid 20th century.

In practice, however, these developments led to a sharp divergence from physics, particularly from 1950 to 1983. Later this was criticised, for example by Vladimir Arnold, as too much Hilbert, not enough Poincaré. The point does not yet seem to be settled, in that string theory pulls one way, while discrete mathematics pulls back towards proof as central.

Applied Mathematics

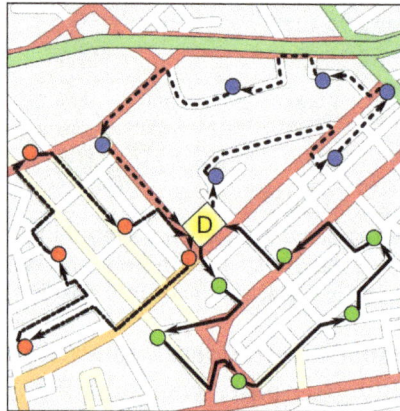

Efficient solutions to the vehicle routing problem require tools from combinatorial optimization and integer programming.

Applied mathematics is the application of mathematical methods by different fields such as science, engineering, business, computer science, and industry. Thus, applied mathematics is a combination of mathematical science and specialized knowledge. The term "applied mathematics" also describes the professional specialty in which mathematicians work on practical problems by formulating and studying mathematical models. In the past, practical applications have motivated the development of mathematical theories, which then became the subject of study in pure mathematics where abstract concepts are studied for their own sake. The activity of applied mathematics is thus intimately connected with research in pure mathematics.

Divisions

Today, the term "applied mathematics" is used in a broader sense. It includes the classical areas noted above as well as other areas that have become increasingly important in applications. Even fields such as number theory that are part of pure mathematics are now important in applications (such as cryptography), though they are not generally considered to be part of the field of applied mathematics *per se*. Sometimes, the term "applicable mathematics" is used to distinguish between the traditional applied mathematics that developed alongside physics and the many areas of mathematics that are applicable to real-world problems today.

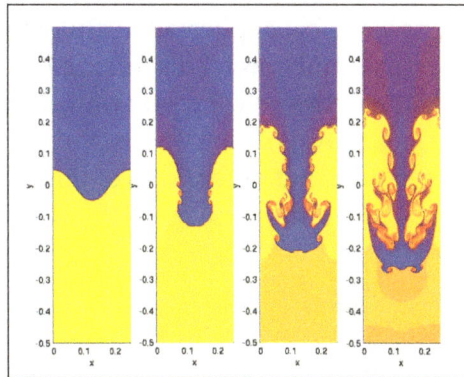

Fluid mechanics is often considered a branch of applied mathematics and mechanical engineering.

There is no consensus as to what the various branches of applied mathematics are. Such categorizations are made difficult by the way mathematics and science change over time, and also by the way universities organize departments, courses, and degrees.

Many mathematicians distinguish between "applied mathematics", which is concerned with mathematical methods, and the "applications of mathematics" within science and engineering. A biologist using a population model and applying known mathematics would not be *doing* applied mathematics, but rather *using* it; however, mathematical biologists have posed problems that have stimulated the growth of pure mathematics. Mathematicians such as Poincaré and Arnold deny the existence of "applied mathematics" and claim that there are only "applications of mathematics." Similarly, non-mathematicians blend applied mathematics and applications of mathematics. The use and development of mathematics to solve industrial problems is also called "industrial mathematics".

The success of modern numerical mathematical methods and software has led to the emergence of computational mathematics, computational science, and computational engineering, which use high-performance computing for the simulation of phenomena and the solution of problems in the sciences and engineering. These are often considered interdisciplinary.

Utility

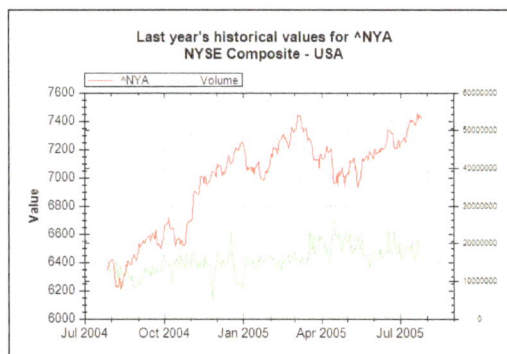

Mathematical finance is concerned with the modelling of financial markets.

Historically, mathematics was most important in the natural sciences and engineering. However, since World War II, fields outside the physical sciences have spawned the creation of new areas of mathematics, such as game theory and social choice theory, which grew out of economic considerations.

The advent of the computer has enabled new applications: studying and using the new computer technology itself (computer science) to study problems arising in other areas of science (computational science) as well as the mathematics of computation (for example, theoretical computer science, computer algebra, numerical analysis). Statistics is probably the most widespread mathematical science used in the social sciences, but other areas of mathematics, most notably economics, are proving increasingly useful in these disciplines.

Status in Academic Departments

Academic institutions are not consistent in the way they group and label courses, programs, and degrees in applied mathematics. At some schools, there is a single mathematics department, whereas others have separate departments for Applied Mathematics and (Pure) Mathematics. It is very common for Statistics departments to be separated at schools with graduate programs, but many undergraduate-only institutions include statistics under the mathematics department.

Many applied mathematics programs (as opposed to departments) consist of primarily cross-listed courses and jointly appointed faculty in departments representing applications. Some Ph.D. programs in applied mathematics require little or no coursework outside mathematics, while others require substantial coursework in a specific area of application. In some respects this difference reflects the distinction between "application of mathematics" and "applied mathematics".

Some universities in the UK host departments of Applied Mathematics and Theoretical Physics, but it is now much less common to have separate departments of pure and applied mathematics. A notable exception to this is the Department of Applied Mathematics and Theoretical Physics at the University of Cambridge, housing the Lucasian Professor of Mathematics whose past holders include Isaac Newton, Charles Babbage, James Lighthill, Paul Dirac and Stephen Hawking.

Schools with separate applied mathematics departments range from Brown University, which has a large Division of Applied Mathematics that offers degrees through the doctorate, to Santa Clara University, which offers only the M.S. in applied mathematics. Research universities dividing their mathematics department into pure and applied sections include MIT. Brigham Young University also has an Applied and Computational Emphasis (ACME), a program that allows student to graduate with a Mathematics degree, with an emphasis in Applied Math. Students in this program also learn another skill (Computer Science, Engineering, Physics, Pure Math, etc.) to supplement their applied math skills.

Associated Mathematical Sciences

Applied mathematics is closely related to other mathematical sciences.

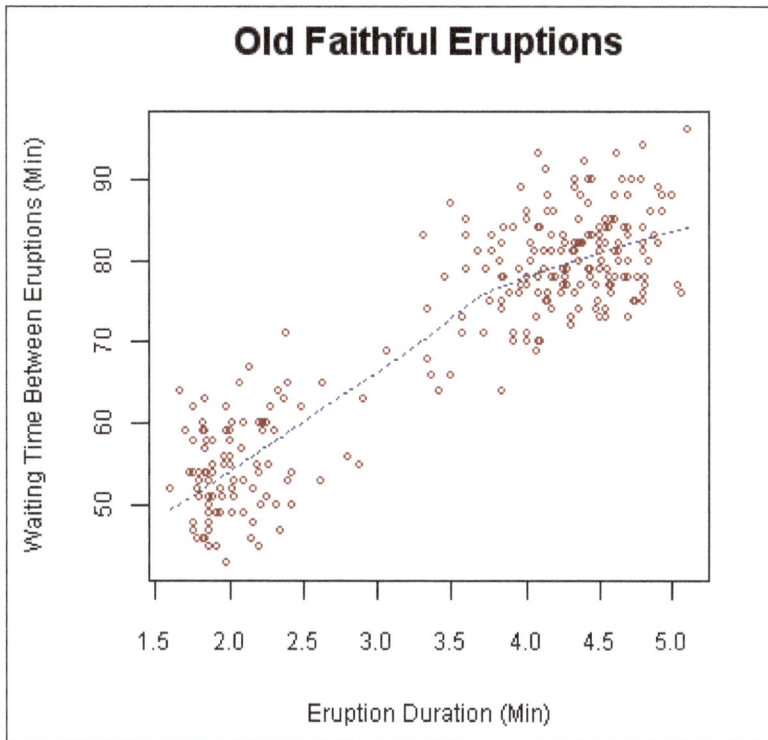

Applied mathematics has substantial overlap with statistics.

Scientific Computing

Scientific computing includes applied mathematics (especially numerical analysis), computing science (especially high-performance computing), and mathematical modelling in a scientific discipline.

Computer Science

Computer science relies on logic, algebra, graph theory, and combinatorics. It entails software engineering.

Operations Research and Management Science

Operations research and management science are often taught in faculties of engineering, business, and public policy.

Statistics

Applied mathematics has substantial overlap with the discipline of statistics. Statistical theorists study and improve statistical procedures with mathematics, and statistical research often raises mathematical questions. Statistical theory relies on probability and decision theory, and makes extensive use of scientific computing, analysis, and optimization; for the design of experiments, statisticians use algebra and combinatorial design. Applied mathematicians and statisticians often work in a department of mathematical sciences (particularly at colleges and small universities).

Actuarial Science

Actuarial science applies probability, statistics, and economic theory to assess risk in insurance, finance and other industries and professions.

Mathematical Economics

Mathematical economics is the application of mathematical methods to represent theories and analyze problems in economics. The applied methods usually refer to nontrivial mathematical techniques or approaches. Mathematical economics is based on statistics, probability, mathematical programming (as well as other computational methods), operations research, game theory, and some methods from mathematical analysis. In this regard, it resembles (but is distinct from) financial mathematics, another part of applied mathematics.

According to the Mathematics Subject Classification (MSC), mathematical economics falls into the Applied mathematics/other classification of category 91: Game theory, economics, social and behavioral sciences with MSC2010 classifications for 'Game theory' at codes 91Axx and for 'Mathematical economics' at codes 91Bxx.

Applicable Mathematics

Applicable mathematics is a subdiscipline of applied mathematics, although there is no consensus as to a precise definition. Sometimes the term "applicable mathematics" is used to distinguish between the traditional applied mathematics that developed alongside physics and the many areas of mathematics that are applicable to real-world problems today.

Mathematicians often distinguish between "applied mathematics" on the one hand, and the "applications of mathematics" or "applicable mathematics" both within and outside of science and engineering, on the other. Some mathematicians emphasize the term applicable mathematics to separate or delineate the traditional applied areas from new applications arising from fields that were previously seen as pure mathematics. For example, from this viewpoint, an ecologist or geographer using population models and applying known mathematics would not be doing applied, but rather applicable, mathematics. Even fields such as number theory that are part of pure mathematics are now important in applications (such as cryptography), though they are not generally considered to be part of the field of applied mathematics *per se*.

Other authors prefer describing applicable mathematics as a union of "new" mathematical applications with the traditional fields of applied mathematics. With this outlook, the terms applied mathematics and applicable mathematics are thus interchangeable.

Other Disciplines

The line between applied mathematics and specific areas of application is often blurred. Many universities teach mathematical and statistical courses outside the respective departments, in departments and areas including business, engineering, physics, chemistry, psychology, biology, computer science, scientific computation, and mathematical physics.

References

- University of strathclyde (17 january 2008), industrial mathematics, archived from the original on 2012-08-04, Retrieved 8 January 2009

- Mathematics: livescience.com, Retrieved 8 January, 2019

- Perspectives on mathematics education: papers submitted by members of the bacomet group, pgs 82-3.editors: h. Christiansen, a.g. Howson, m. Otte. Volume 2 of mathematics education library; springer science & business media, 2012. Isbn 9400945043, 9789400945043

- Branches-of-mathematics: shriramrahatgaonkar.weebly.com, Retrieved 9 February, 2019

- Santa clara university dept of applied mathematics, archived from the original on 2011-05-04, Retrieved 2011-03-05

Arithmetic 2

- **Number Theory**

- **Addition**

- **Subtraction**

- **Multiplication**

- **Division**

- **Decimal**

Arithmetic is the branch of mathematics which refers to the study of properties and manipulation of numbers. Number theory, addition, subtraction, multiplication, division, decimal, etc. are some the concepts that fall under its domain. All the diverse concepts of arithmetic have been carefully analyzed in this chapter.

Arithmetic is the branch of mathematics in which numbers, relations among numbers, and observations on numbers are studied and used to solve problems.

Arithmetic refers generally to the elementary aspects of the theory of numbers, arts of mensuration (measurement), and numerical computation (that is, the processes of addition, subtraction, multiplication, division, raising to powers, and extraction of roots). Its meaning, however, has not been uniform in mathematical usage. An eminent German mathematician, Carl Friedrich Gauss, in Disquisitiones Arithmeticae, and certain modern-day mathematicians have used the term to include more advanced topics.

Fundamental Definitions and Laws

Natural Numbers

In a collection (or set) of objects (or elements), the act of determining the number of objects present is called counting. The numbers thus obtained are called the counting numbers or natural numbers (1, 2, 3, ...). For an empty set, no object is present, and the count yields the number 0, which, appended to the natural numbers, produces what are known as the whole numbers.

If objects from two sets can be matched in such a way that every element from each set is uniquely paired with an element from the other set, the sets are said to be equal or equivalent. The concept of equivalent sets is basic to the foundations of modern mathematics and has been introduced into primary education, notably as part of the "new math" that has been alternately acclaimed and decried since it appeared in the 1960s.

Addition and Multiplication

Combining two sets of objects together, which contain a and b elements, a new set is formed that contains a + b = c objects. The number c is called the sum of a and b; and each of the latter is called a summand. The operation of forming the sum is called addition, the symbol + being read as "plus." This is the simplest binary operation, where binary refers to the process of combining two objects.

From the definition of counting it is evident that the order of the summands can be changed and the order of the operation of addition can be changed, when applied to three summands, without affecting the sum. These are called the commutative law of addition and the associative law of addition, respectively.

If there exists a natural number k such that $a = b + k$, it is said that a is greater than b (written $a > b$) and that b is less than a (written $b < a$). If a and b are any two natural numbers, then it is the case that either $a = b$ or $a > b$ or $a < b$ (the trichotomy law).

From the above laws, it is evident that a repeated sum such as 5 + 5 + 5 is independent of the way in which the summands are grouped; it can be written 3 × 5. Thus, a second binary operation called multiplication is defined. The number 5 is called the multiplicand; the number 3, which denotes the number of summands, is called the multiplier; and the result 3 × 5 is called the product. The symbol × of this operation is read "times." If such letters as a and b are used to denote the numbers, the product $a \times b$ is often written $a.b$ or simply ab.

If three rows of five dots each are written, as illustrated below:

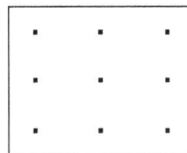

It is clear that the total number of dots in the array is 3 × 5, or 15. This same number of dots can evidently be written in five rows of three dots each, whence 5 × 3 = 15. The argument is general, leading to the law that the order of the multiplicands does not affect the product, called the commutative law of multiplication. But it is notable that this law does not apply to all mathematical entities. Indeed, much of the mathematical formulation of modern physics, for example, depends crucially on the fact that some entities do not commute.

By the use of a three-dimensional array of dots, it becomes evident that the order of multiplication when applied to three numbers does not affect the product. Such a law is called the associative law of multiplication. If the 15 dots written above are separated into two sets, as shown,

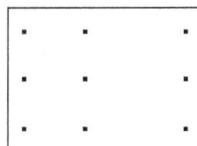

then the first set consists of three columns of three dots each, or 3 × 3 dots; the second set consists of two columns of three dots each, or 2 × 3 dots; the sum (3 × 3) + (2 × 3) consists of 3 + 2 = 5 columns of three dots each, or (3 + 2) × 3 dots. In general, one may prove that the multiplication of a sum by a number is the same as the sum of two appropriate products. Such a law is called the distributive law.

Integers

Subtraction has not been introduced for the simple reason that it can be defined as the inverse of addition. Thus, the difference $a - b$ of two numbers a and b is defined as a solution x of the equation $b + x = a$. If a number system is restricted to the natural numbers, differences need not always exist, but, if they do, the five basic laws of arithmetic, as already discussed, can be used to prove that they are unique. Furthermore, the laws of operations of addition and multiplication can be extended to apply to differences. The whole numbers (including zero) can be extended to include the solution of $1 + x = 0$, that is, the number -1, as well as all products of the form $-1 \times n$, in which n is a whole number. The extended collection of numbers is called the integers, of which the positive integers are the same as the natural numbers. The numbers that are newly introduced in this way are called negative integers.

Exponents

Just as a repeated sum $a + a + \cdots + a$ of k summands is written ka, so a repeated product $a \times a \times \cdots \times a$ of k factors is written a^k. The number k is called the exponent, and a the base of the power a^k.

The fundamental laws of exponents follow easily from the definitions, and other laws are immediate consequences of the fundamental ones.

Fundamental laws of exponents		
Products:	$b^m b^n$	$= b^{m+n}$
Ratios:	$\dfrac{b^m}{b^n}$	$= b^{m-n}$
Powers:	$(b^m)^n$	$= b^{mn}$
Roots:	$\sqrt[q]{b^n}$	$= b^{n/q}$

Theory of Divisors

At this point an interesting development occurs, for, so long as only additions and multiplications are performed with integers, the resulting numbers are invariably themselves integers—that is, numbers of the same kind as their antecedents. This characteristic changes drastically, however, as soon as division is introduced. Performing division (its symbol ÷, read "divided by") leads to results, called quotients or fractions, which surprisingly include numbers of a new kind—namely, rationals—that are not integers. These, though arising from the combination of integers, patently constitute a distinct extension of the natural-number and integer concepts as defined above. By means of the application of the division operation, the domain of the natural numbers becomes extended and enriched immeasurably beyond the integers.

The preceding illustrates one of the proclivities that are often associated with mathematical thought: relatively simple concepts (such as integers), initially based on very concrete operations (for example, counting), are found to be capable of assuming novel meanings and potential uses, extending far beyond the limits of the concept as originally defined. A similar extension of basic concepts, with even more powerful results, will be found with the introduction of irrationals.

A second example of this pattern is presented by the following: Under the primitive definition of exponents, with k equal to either zero or a fraction, a^k would, at first sight, appear to be utterly devoid of meaning. Clarification is needed before writing a repeated product of either zero factors or a fractional number of factors. Considering the case $k = 0$, a little reflection shows that a^0 can, in fact, assume a perfectly precise meaning, coupled with an additional and quite extraordinary property. Since the result of dividing any (nonzero) number by itself is 1, or unity, it follows that,

$$a^m \div a^m = a^{m-m} = a^0 = 1.$$

Not only can the definition of a^k be extended to include the case $k = 0$, but the ensuing result also possesses the noteworthy property that it is independent of the particular (nonzero) value of the base a. A similar argument may be given to show that a^k is a meaningful expression even when k is negative, namely,

$$a^{-k} = 1 / a^k.$$

The original concept of exponent is thus broadened to a great extent.

Fundamental Theory

If three positive integers a, b, and c are in the relation $ab = c$, it is said that a and b are divisors or factors of c, or that a divides c (written $a \mid c$), and b divides c. The number c is said to be a multiple of a and a multiple of b.

The number 1 is called the unit, and it is clear that 1 is a divisor of every positive integer. If c can be expressed as a product ab in which a and b are positive integers each greater than 1, then c is called composite. A positive integer neither 1 nor composite is called a prime number. Thus, 2, 3, 5, 7, 11, 13, 17, 19, ... are prime numbers. The ancient Greek mathematician Euclid proved in his Elements that there are infinitely many prime numbers.

The fundamental theorem of arithmetic was proved by Gauss in his Disquisitiones Arithmeticae. It states that every composite number can be expressed as a product of prime numbers and that, save for the order in which the factors are written, this representation is unique. Gauss's theorem follows rather directly from another theorem of Euclid to the effect that if a prime divides a product, then it also divides one of the factors in the product; for this reason the fundamental theorem is sometimes credited to Euclid.

For every finite set a_1, a_2,..., a_k of positive integers, there exists a largest integer that divides each of these numbers, called their greatest common divisor (GCD). If the GCD = 1, the numbers are said to be relatively prime. There also exists a smallest positive integer that is a multiple of each of the numbers, called their least common multiple (LCM).

A systematic method for obtaining the GCD and LCM starts by factoring each (*where i* = 1, 2, ..., *k*) into a product of primes $p_1, p_2, \ldots, p_h$, with the number of times that each distinct prime occurs indicated by qi; thus,

$$a_i = p_1^{q1} p_2^{q2} \cdots p_h^{qh}$$

Then the GCD is obtained by multiplying together each prime that occurs in every a_i as many times as it occurs the fewest (smallest power) among all of the a_i. The LCM is obtained by multiplying together each prime that occurs in any of the a_i as many times as it occurs the most (largest power) among all of the a_i. An example is easily constructed. Given $a_1 = 3,000 = 2^3 \times 3^1 \times 5^3$ and $a_2 = 2,646 = 2^1 \times 3^3 \times 7^2$, the $GCD = 2^1 \times 3^1 = 6$ and the $LCM = 2^3 \times 3^3 \times 5^3 \times 7^2 = 1,323,000$. When only two numbers are involved, the product of the GCD and the LCM equals the product of the original numbers.

Some Divisibility Rules

Divisor	Condition
1	The number is even.
2	The sum of the digits in the number is divisible by 3.
3	The last two digits in the number form a number that is divisible by 4.
4	The number ends in 0 or 5.
5	The number is even and the sum of its digits is divisible by 3.
6	The last three digits in the number form a number that is divisible by 8.
7	The sum of the digits in the number is divisible by 9.
8	The number ends in 0.
9	The difference between the sum of the number's digits in the odd places and that of the digits in the even places is either 0 or divisible by 11.

If a and b are two positive integers, with $a > b$, two whole numbers q and r exist such that $a = qb + r$, with r less than b. The number q is called the partial quotient (the quotient if $r = 0$), and r is called the remainder. Using a process known as the Euclidean algorithm, which works because the GCD of a and b is equal to the GCD of b and r, the GCD can be obtained without first factoring the numbers a and b into prime factors. The Euclidean algorithm begins by determining the values of q and r, after which b and r assume the role of a and b and the process repeats until finally the remainder is zero; the last positive remainder is the GCD of the original two numbers. For example, starting with 544 and 119:

- 544 = 4 × 119 + 68;
- 119 = 1 × 68 + 51;
- 68 = 1 × 51 + 17;
- 51 = 3 × 17.

Thus, the GCD of 544 and 119 is 17.

Rational Numbers

From a less abstract point of view, the notion of division, or of fraction, may also be considered to arise as follows: if the duration of a given process is required to be known to an accuracy of better than one hour, the number of minutes may be specified; or, if the hour is to be retained as the fundamental unit, each minute may be represented by 1/60 or by $\frac{1}{60}$.

In general, the fractional unit $1/d$ is defined by the property d $\times$ $1/d = 1$. The number $n \times 1/d$ is written n/d and is called a common fraction. It may be considered as the quotient of n divided by d. The number d is called the denominator (it determines the fractional unit or denomination), and n is called the numerator (it enumerates the number of fractional units that are taken). The numerator and denominator together are called the terms of the fraction. A positive fraction n/d is said to be proper if $n < d$; otherwise it is improper.

The numerator and denominator of a fraction are not unique, since for every positive integer k, the numerator and denominator of a fraction can each simultaneously be multiplied by the integer k without altering the fractional value. Every fraction can be written as the quotient of two relatively prime integers, however. In this form it is said to be in lowest terms.

The integers and fractions constitute what are called the rational numbers. The five fundamental laws stated earlier with regard to the positive integers can be generalized to apply to all rational numbers.

Adding and Subtracting Fractions

From the definition of fraction it follows that the sum (or difference) of two fractions having the same denominator is another fraction with this denominator, the numerator of which is the sum (or difference) of the numerators of the given fractions. Two fractions having different denominators may be added or subtracted by first reducing them to fractions with the same denominator. Thus, to add a/b and c/d, the LCM of b and d, often called the least common denominator of the fractions, must be determined. It follows that there exist numbers k and l such that $kb = ld$, and both fractions can be written with this common denominator, so that the sum or difference of the fractions is obtained by the simple operation of adding or subtracting the new numerators and placing the value over the new denominator.

Multiplying and Dividing Fractions

In order to multiply two fractions—in case one of the numbers is a whole number, it is placed over the number 1 to create a fraction—the numerators and denominators are multiplied separately to produce the new fraction's numerator and denominator: $a/b \times c/d = ac/bd$. In order to divide by a fraction, it must be inverted—that is, the numerator and denominator interchanged—after which it becomes a multiplication problem: $a/b \div c/d = a/b \times d/c = ad/bc$.

Theory of Rationals

A method of introducing the positive rational numbers that is free from intuition (that is, with all logical steps included) was given in 1910 by the German mathematician Ernst Steinitz. In considering the set of all number pairs (a, b), (c, d), ...in which a, b, c, d, ...are positive integers, the equals relation $(a, b) = (c, d)$ is defined to mean that $ad = bc$, and the two operations $+$ and $\times$ are defined so that the sum of a pair $(a, b) + (c, d) = (ad + bc, bd)$ is a pair and the product of a pair $(a, b) \times (c, d) = (ac, bd)$ is a pair. It can be proved that, if these sums and products are properly specified, the fundamental laws of arithmetic hold for these pairs and that the pairs of the type $(a, 1)$ are abstractly identical with the positive integers a. Moreover, $b \times (a, b) = a$, so that the pair (a, b) is abstractly identical with the fraction a/b.

Irrational Numbers

It was known to the Pythagoreans that, given a straight line segment a and a unit segment u, it is not always possible to find a fractional unit such that both a and u are multiples of it. For instance, if the sides of an isosceles right triangle have length 1, then by the Pythagorean theorem the hypotenuse has a length the square of which must be 2. But there exists no rational number the square of which is 2.

Eudoxus of Cnidus, a contemporary of Plato, established the technique necessary to extend numbers beyond the rationals. His contribution, one of the most important in the history of mathematics, was included in Euclid's Elements and elsewhere, and then it lay dormant until the modern period of growth in mathematical analysis in Germany in the 19th century.

It is customary to assume on an intuitive basis that, corresponding to every line segment and every unit length, there exists a number (called a positive real number) that represents the length of the line segment. Not all such numbers are rational, but every one can be approximated arbitrarily closely by a rational number. That is, if x is a positive real number and ε is any positive rational number—no matter how small—it is possible to find two positive rational numbers a and b within ε distance from each other such that x is between them; in symbols, given any $\varepsilon > 0$, there exist positive rational numbers a and b such that $b - a < \varepsilon$ and $a < x < b$. In problems in mensuration, irrational numbers are usually replaced by suitable rational approximations.

A rigorous development of the irrational numbers is beyond the scope of arithmetic. They are most satisfactorily introduced by means of Dedekind cuts, as introduced by the German mathematician Richard Dedekind, or sequences of rationals, as introduced by Eudoxus and developed by the German mathematician Georg Cantor. These methods are discussed in analysis.

The employment of irrational numbers greatly increases the scope and usefulness of arithmetic. For instance, if n is any whole number and a is any positive real number, there exists a unique positive real number $n\sqrt{a}$, called the nth root of a, whose nth power is a. The root symbol $\sqrt{}$ is a conventionalized r for radix, or "root." The term evolution is sometimes applied to the process of finding a rational approximation to an nth root.

Number Theory

Number theory is the branch of mathematics concerned with properties of the positive integers (1, 2, 3, ...). Sometimes called "higher arithmetic," it is among the oldest and most natural of mathematical pursuits.

Number theory has always fascinated amateurs as well as professional mathematicians. In contrast to other branches of mathematics, many of the problems and theorems of number theory can be understood by laypersons, although solutions to the problems and proofs of the theorems often require a sophisticated mathematical background.

Until the mid-20th century, number theory was considered the purest branch of mathematics, with no direct applications to the real world. The advent of digital computers and digital communications revealed that number theory could provide unexpected answers to real-world problems. At the same time, improvements in computer technology enabled number theorists to make remarkable advances in factoring large numbers, determining primes, testing conjectures, and solving numerical problems once considered out of reach.

Modern number theory is a broad subject that is classified into subheadings such as elementary number theory, algebraic number theory, analytic number theory, geometric number theory, and probabilistic number theory. These categories reflect the methods used to address problems concerning the integers.

From Prehistory through Classical Greece

The ability to count dates back to prehistoric times. This is evident from archaeological artifacts, such as a 10,000-year-old bone from the Congo region of Africa with tally marks scratched upon it—signs of an unknown ancestor counting something. Very near the dawn of civilization, people had grasped the idea of "multiplicity" and thereby had taken the first steps toward a study of numbers.

It is certain that an understanding of numbers existed in ancient Mesopotamia, Egypt, China, and India, for tablets, papyri, and temple carvings from these early cultures have survived. A Babylonian tablet known as Plimpton 322 is a case in point. In modern notation, it displays number triples x, y and z with the property that $x^2 + y^2 = z^2$. One such triple is 2,291, 2,700, and 3,541, where $2,291^2 + 2,700^2 = 3,541^2$. This certainly reveals a degree of number theoretic sophistication in ancient Babylon.

Despite such isolated results, a general theory of numbers was nonexistent. For this—as with so much of theoretical mathematics—one must look to the Classical Greeks, whose groundbreaking achievements displayed an odd fusion of the mystical tendencies of the Pythagoreans and the severe logic of Euclid's Elements.

Pythagoras

According to tradition, Pythagoras worked in southern Italy amid devoted followers. His philosophy enshrined number as the unifying concept necessary for understanding everything from planetary motion to musical harmony. Given this viewpoint, it is not surprising that the Pythagoreans attributed quasi-rational properties to certain numbers.

For instance, they attached significance to perfect numbers—i.e., those that equal the sum of their proper divisors. Examples are 6 (whose proper divisors 1, 2, and 3 sum to 6) and 28 (1 + 2 + 4 + 7 + 14). The Greek philosopher Nicomachus of Gerasa (flourished c. AD 100), writing centuries after Pythagoras but clearly in his philosophical debt, stated that perfect numbers represented "virtues, wealth, moderation, propriety, and beauty." (Some modern writers label such nonsense numerical theology).

In a similar vein, the Greeks called a pair of integers amicable ("friendly") if each was the sum of the proper divisors of the other. They knew only a single amicable pair: 220 and 284. One can easily check that the sum of the proper divisors of 284 is 1 + 2 + 4 + 71 + 142 = 220 and the sum of the proper divisors of 220 is 1 + 2 + 4 + 5 + 10 + 11 + 20 + 22 + 44 + 55 + 110 = 284. For those prone to number mysticism, such a phenomenon must have seemed like magic.

Euclid

By contrast, Euclid presented number theory without the flourishes. He began Book VII of his Elements by defining a number as "a multitude composed of units." The plural here excluded 1; for Euclid, 2 was the smallest "number." He later defined a prime as a number "measured by a unit alone" (i.e., whose only proper divisor is 1), a composite as a number that is not prime, and a perfect number as one that equals the sum of its "parts" (i.e., its proper divisors).

From there, Euclid proved a sequence of theorems that marks the beginning of number theory as a mathematical (as opposed to a numerological) enterprise. Four Euclidean propositions deserve special mention.

The first, Proposition 2 of Book VII, is a procedure for finding the greatest common divisor of two whole numbers. This fundamental result is now called the Euclidean algorithm in his honour.

Second, Euclid gave a version of what is known as the unique factorization theorem or the fundamental theorem of arithmetic. This says that any whole number can be factored into the product of primes in one and only one way. For example, 1,960 = 2 × 2 × 2 × 5 × 7 × 7 is a decomposition into prime factors, and no other such decomposition exists. Euclid's discussion of unique factorization is not satisfactory by modern standards, but its essence can be found in Proposition 32 of Book VII and Proposition 14 of Book IX.

Third, Euclid showed that no finite collection of primes contains them all. His argument, Proposition 20 of Book IX, remains one of the most elegant proofs in all of mathematics. Beginning with any finite collection of primes—say, a, b, c, ..., n—Euclid considered the number formed by adding one to their product: $N = (abc\cdots n) + 1$. He then examined the two alternatives:

(1) If N is prime, then it is a new prime not among a, b, c, ..., n because it is larger than all of these. For example, if the original primes were 2, 3, and 7, then $N = (2 \times 3 \times 7) + 1 = 43$ is a larger prime. (2) Alternately, if N is composite, it must have a prime factor which, as Euclid demonstrated, cannot be one of the originals. To illustrate, begin with primes 2, 7, and 11, so that $N = (2 \times 7 \times 11) + 1 = 155$. This is composite, but its prime factors 5 and 31 do not appear among the originals. Either way, a finite set of primes can always be augmented. It follows, by this beautiful piece of logic, that the collection of primes is infinite.

Fourth, Euclid ended Book IX with a blockbuster: if the series $1 + 2 + 4 + 8 + ... + 2^k$ sums to a prime, then the number $N = 2^k \left(1 + 2 + 4 + ... + 2^k\right)$ must be perfect. For example, $1 + 2 + 4 = 7$, a prime, so $4(1 + 2 + 4) = 28$ is perfect. Euclid's "recipe" for perfect numbers was a most impressive achievement for its day.

Diophantus

Of later Greek mathematicians, especially noteworthy is Diophantus of Alexandria, author of Arithmetica. This book features a host of problems, the most significant of which have come to be called Diophantine equations. These are equations whose solutions must be whole numbers. For example, Diophantus asked for two numbers, one a square and the other a cube, such that the sum of their squares is itself a square. In modern symbols, he sought integers x, y, and z such that $\left(x^2\right)^2 + \left(y^3\right)^2 = z^2$. It is easy to find real numbers satisfying this relationship (e.g., $x = \sqrt{2}$, $y = 1$, $and\ z = \sqrt{5}$), but the requirement that solutions be integers makes the problem more difficult. (One answer is $x = 6$, $y = 3$, $and\ z = 45$). Diophantus's work strongly influenced later mathematics.

Number Theory in the East

The millennium following the decline of Rome saw no significant European advances, but Chinese and Indian scholars were making their own contributions to the theory of numbers. Motivated by questions of astronomy and the calendar, the Chinese mathematician Sun Zi tackled multiple Diophantine equations. As one example, he asked for a whole number that when divided by 3 leaves a remainder of 2, when divided by 5 leaves a remainder of 3, and when divided by 7 leaves a remainder of 2 (his answer: 23). Almost a thousand years later, Qin Jiushao gave a general procedure, now known as the Chinese remainder theorem, for solving problems of this sort.

Meanwhile, Indian mathematicians were hard at work. In the 7th century Brahmagupta took up what is now (erroneously) called the Pell equation. He posed the challenge to find a perfect square that, when multiplied by 92 and increased by 1, yields another perfect square. That is, he sought whole numbers x and y such that $92x^2 + 1 = y^2 - a$. Diophantine equation with quadratic terms. Brahmagupta suggested that anyone who could solve this problem within a year earned the right to be called a mathematician. His solution was $x = 120$ and $y = 1,151$.

In addition, Indian scholars developed the so-called Hindu-Arabic numerals—the base-10 notation subsequently adopted by the world's mathematical and civil communities. Although more number representation than number theory, these numerals have prevailed due to their simplicity and ease of use. The Indians employed this system—including the zero—as early as AD 800.

At about this time, the Islamic world became a mathematical powerhouse. Situated on trade routes between East and West, Islamic scholars absorbed the works of other civilizations and augmented these with homegrown achievements. For example, Thabit ibn Qurrah (active in Baghdad in the 9th century) returned to the Greek problem of amicable numbers and discovered a second pair: 17,296 and 18,416.

Modern Number Theory

As mathematics filtered from the Islamic world to Renaissance Europe, number theory received little serious attention. The period from 1400 to 1650 saw important advances in geometry, algebra, and probability, not to mention the discovery of both logarithms and analytic geometry. But number theory was regarded as a minor subject, largely of recreational interest.

Pierre de Fermat

Credit for changing this perception goes to Pierre de Fermat, a French magistrate with time on his hands and a passion for numbers. Although he published little, Fermat posed the questions and identified the issues that have shaped number theory ever since. Here are a few examples:

- In 1640 he stated what is known as Fermat's little theorem—namely, that if p is prime and a is any whole number, then p divides evenly into $a^p - a$. Thus, if $p = 7$ and $a = 12$, the far-from-obvious conclusion is that 7 is a divisor of $12^7 - 12 = 35{,}831{,}796$. This theorem is one of the great tools of modern number theory.

- Fermat investigated the two types of odd primes: those that are one more than a multiple of 4 and those that are one less. These are designated as the 4k + 1 primes and the 4k − 1 primes, respectively. Among the former are $5 = 4 \times 1 + 1$ and $97 = 4 \times 24 + 1$; among the latter are $3 = 4 \times 1 - 1$ and $79 = 4 \times 20 - 1$. Fermat asserted that any prime of the form $4k + 1$ can be written as the sum of two squares in one and only one way, whereas a prime of the form $4k - 1$ cannot be written as the sum of two squares in any manner whatever. Thus, $5 = 2^2 + 1^2$ and $97 = 9^2 + 4^2$, and these have no alternative decompositions into sums of squares. On the other hand, 3 and 79 cannot be so decomposed. This dichotomy among primes ranks as one of the landmarks of number theory.

- In 1638 Fermat asserted that every whole number can be expressed as the sum of four or fewer squares. He claimed to have a proof but did not share it.

- Fermat stated that there cannot be a right triangle with sides of integer length whose area is a perfect square. This amounts to saying that there do not exist integers x, y, z, and w such that x² + y² = z² (the Pythagorean relationship) and that $w^2 = 1/2 (base)(height) = xy / 2$.

Uncharacteristically, Fermat provided a proof of this last result. He used a technique called infinite descent that was ideal for demonstrating impossibility. The logical strategy assumes that there are whole numbers satisfying the condition in question and then generates smaller whole numbers satisfying it as well. Reapplying the argument over and over, Fermat produced an endless sequence of decreasing whole numbers. But this is impossible, for any set of positive integers must contain a smallest member. By this contradiction, Fermat concluded that no such numbers can exist in the first place.

Two other assertions of Fermat should be mentioned. One was that any number of the form $2^{2^n} + 1$ must be prime. He was correct if $n = $ 0, 1, 2, 3 and 4, for the formula yields primes $2^{2^0} + 1 = 3$, $2^{2^1} + 1 = 5$, $2^{2^2} + 1 = 17$, $2^{2^3} + 1 = 257$, and $2^{2^4} + 1 = 65{,}537$. These are now called Fermat primes. Unfortunately for his reputation, the next such number $2^{2^5} + 1 = 2^{32} + 1 = 4{,}294{,}967{,}297$ is not a prime (more about that later). Even Fermat was not invincible.

The second assertion is one of the most famous statements from the history of mathematics. While reading Diophantus's Arithmetica, Fermat wrote in the book's margin: "To divide a cube into two cubes, a fourth power, or in general any power whatever into two powers of the same denomination above the second is impossible." He added that "I have assuredly found an admirable proof of this, but the margin is too narrow to contain it."

In symbols, he was claiming that if $n > 2,$ there are no whole numbers x, y, z such that $x^n + y^n = z^n$, a statement that came to be known as Fermat's last theorem. For three and a half centuries, it defeated all who attacked it, earning a reputation as the most famous unsolved problem in mathematics.

Despite Fermat's genius, number theory still was relatively neglected. His reluctance to supply proofs was partly to blame, but perhaps more detrimental was the appearance of the calculus in the last decades of the 17th century. Calculus is the most useful mathematical tool of all, and scholars eagerly applied its ideas to a range of real-world problems. By contrast, number theory seemed too "pure," too divorced from the concerns of physicists, astronomers, and engineers.

Number Theory in the 18th Century

Credit for bringing number theory into the mainstream, for finally realizing Fermat's dream, is due to the 18th century's dominant mathematical figure, the Swiss Leonhard Euler. Euler was the most prolific mathematician ever—and one of the most influential—and when he turned his attention to number theory, the subject could no longer be ignored.

Initially, Euler shared the widespread indifference of his colleagues, but he was in correspondence with Christian Goldbach, a number theory enthusiast acquainted with Fermat's work. Like an insistent salesman, Goldbach tried to interest Euler in the theory of numbers, and eventually his insistence paid off.

It was a letter of December 1, 1729, in which Goldbach asked Euler, "Is Fermat's observation known to you, that all numbers $2^{2^n} + 1$ are primes?" This caught Euler's attention. Indeed, he showed that Fermat's assertion was wrong by splitting the number $2^{2^5} + 1$ into the product of 641 and 6,700,417.

Through the next five decades, Euler published over a thousand pages of research on number theory, much of it furnishing proofs of Fermat's assertions. In 1736 he proved Fermat's little theorem. By mid-century he had established Fermat's theorem that primes of the form $4k + 1$ can be uniquely expressed as the sum of two squares. He later took up the matter of perfect numbers, demonstrating that any even perfect number must assume the form discovered by Euclid 20 centuries earlier. And when he turned his attention to amicable numbers—of which, by this time, only three pairs were known—Euler vastly increased the world's supply by finding 58 new ones.

Of course, even Euler could not solve every problem. He gave proofs, or near-proofs, of Fermat's last theorem for exponents $n = 3$ and $n = 4$ but despaired of finding a general solution. And he was completely stumped by Goldbach's assertion that any even number greater than 2 can be written as the sum of two primes. Euler endorsed the result—today known as the Goldbach conjecture—but acknowledged his inability to prove it.

Euler gave number theory a mathematical legitimacy, and thereafter progress was rapid. In 1770, for instance, Joseph-Louis Lagrange proved Fermat's assertion that every whole number can be written as the sum of four or fewer squares. Soon thereafter, he established a beautiful result known as Wilson's theorem: p is prime if and only if p divides evenly into:

$$[(p-1) \times (p-2) \times \cdots \times 3 \times 2 \times 1] + 1.$$

Number Theory in the 19th Century

Disquisitiones Arithmeticae

Of immense significance was the 1801 publication of Disquisitiones Arithmeticae by Carl Friedrich Gauss. This became, in a sense, the holy writ of number theory. In it Gauss organized and summarized much of the work of his predecessors before moving boldly to the frontier of research. Observing that the problem of resolving composite numbers into prime factors is "one of the most important and useful in arithmetic," Gauss provided the first modern proof of the unique factorization theorem. He also gave the first proof of the law of quadratic reciprocity, a deep result previously glimpsed by Euler. To expedite his work, Gauss introduced the idea of congruence among numbers—i.e., he defined a and b to be congruent modulo m (written a ≡ b mod m) if m divides evenly into the difference a − b. For instance, 39 ≡ 4 mod 7. This innovation, when combined with results like Fermat's little theorem, has become an indispensable fixture of number theory.

From Classical to Analytic Number Theory

Inspired by Gauss, other 19th-century mathematicians took up the challenge. Sophie Germain, who once stated, "I have never ceased thinking about the theory of numbers," made important contributions to Fermat's last theorem, and Adrien-Marie Legendre and Peter Gustav Lejeune Dirichlet confirmed the theorem for $n = 5$ — i.e. they showed that the sum of two fifth powers cannot be a fifth power. In 1847 Ernst Kummer went further, demonstrating that Fermat's last theorem was true for a large class of exponents; unfortunately, he could not rule out the possibility that it was false for a large class of exponents, so the problem remained unresolved.

The same Dirichlet (who reportedly kept a copy of Gauss's Disquisitiones Arithmeticae by his bedside for evening reading) made a profound contribution by proving that, if a and b have no common factor, then the arithmetic progression a, $a + b$, $a + 2b$, $a + 3b$, ... must contain infinitely many primes. Among other things, this established that there are infinitely many $4k + 1$ primes and infinitely many $4k - 1$ primes as well. But what made this theorem so exceptional was Dirichlet's method of proof: he employed the techniques of calculus to establish a result in number theory. This surprising but ingenious strategy marked the beginning of a new branch of the subject: analytic number theory.

Prime Number Theorem

One of the supreme achievements of 19th-century mathematics was the prime number theorem, and it is worth a brief digression. To begin, designate the number of primes less than or equal to n by $\pi(n)$. Thus $\pi(10) = 4$ because 2, 3, 5, and 7 are the four primes not exceeding 10. Similarly $\pi(25) = 9$ and $\pi(100) = 25$. Next, consider the proportion of numbers less than or equal to n that are prime—i.e., $\pi(n)/n$. Clearly $\pi(10)/10 = 0.40$, meaning that 40 percent of the numbers not exceeding 10 are prime. Other proportions are shown in the table:

Prime number theorem (illustrated by selected values n from 10^2 to 10^{14})				
n	$\pi(n) =$ number of primes less than or equal to n	$\dfrac{\pi(n)}{n} =$ proportion of primes among the first n numbers	$\dfrac{1}{\log n} =$ predicted proportion of primes among the first n numbers	
10^2	25	0.2500	0.2172	
10^4	1,229	0.1229	0.1086	
10^6	78,498	0.0785	0.0724	
10^8	5,761,455	0.0570	0.0543	
10^{10}	455,052,511	0.0455	0.0434	
10^{12}	37,607,912,018	0.0377	0.0362	
10^{14}	3,204,941,750,802	0.0320	0.0310	

A pattern is anything but clear, but the prime number theorem identifies one, at least approximately, and thereby provides a rule for the distribution of primes among the whole numbers. The theorem says that, for large n, the proportion $\pi(n)/n$ is roughly $1/\log n$, where $\log n$ is the natural logarithm of n. This link between primes and logs is nothing short of extraordinary.

One of the first to perceive this was the young Gauss, whose examination of log tables and prime numbers suggested it to his fertile mind. Following Dirichlet's exploitation of analytic techniques in number theory, Bernhard Riemann and Pafnuty Chebyshev made substantial progress before the prime number theorem was proved in 1896 by Jacques Hadamard and Charles Jean de la Vallée-Poussin. This brought the 19th century to a triumphant close.

Number Theory in the 20th Century

The next century saw an explosion in number theoretic research. Along with classical and analytic number theory, scholars now explored specialized subfields such as algebraic number theory, geometric number theory, and combinatorial number theory. The concepts became more abstract and the techniques more sophisticated. Unquestionably, the subject had grown beyond Fermat's wildest dreams.

One of the great contributors from early in the 20th century was the incandescent genius Srinivasa Ramanujan. Ramanujan, whose formal training was as limited as his life was short, burst upon the mathematical scene with a series of brilliant discoveries. Analytic number theory was among his specialties, and his publications carried titles such as "Highly composite numbers" and "Proof that almost all numbers n are composed of about log(log n) prime factors."

A legendary figure in 20th-century number theory was Paul Erdős, a Hungarian genius known for his deep insights, his vast circle of collaborators, and his personal eccentricities. At age 18, Erdős published a much-simplified proof of a theorem of Chebyshev stating that, if $n \geq 2$, then there must be a prime between n and $2n$. This was the first in a string of number theoretic results that would span most of the century. In the process, Erdős—who also worked in combinatorics, graph theory, and dimension theory—published over 1,500 papers with more than 500 collaborators from around the world. He achieved this astonishing output while living more or less out of a suitcase, traveling constantly from one university to another in pursuit of new mathematics. It was not uncommon for him to arrive, unannounced, with the declaration that "My brain is open" and then to plunge into the latest problem with gusto.

Two later developments deserve mention. One was the invention of the electronic computer, whose speed has been advantageously applied to number theoretic questions. As an example, Euler once speculated that at least four fourth powers must be added together for the sum to be a fourth power. But in 1988, using a combination of mathematical insight and computer muscle, the American Noam Elkies discovered that $2,682,440^4 + 15,365,639^4 + 18,796,760^4 = 20,615,673^4$ —a stupendous counterexample that destroyed Euler's conjecture. The number on the right contains 30 digits, so there is little wonder that Euler missed it.

Second, number theory acquired an applied flavour, for it became instrumental in designing encryption schemes widely used in government and business. These rely upon the factorization of gigantic numbers into primes—a factorization that the code's user knows and the potential code-breaker does not. This application runs counter to the long-held perception of number theory as beautiful but essentially useless.

Twentieth-century number theory reached a much-publicized climax in 1995, when Fermat's last theorem was proved by the Englishman Andrew Wiles, with timely assistance from his British colleague Richard Taylor. Wiles succeeded where so many had failed with a 130-page proof of incredible complexity, one that certainly would not fit into any margin.

Unsolved Problems

This triumph notwithstanding, number theory remains the source of many unsolved problems, some of the most perplexing of which sound innocent enough. For example:

- Do any odd perfect numbers exist?

- Are there infinitely many primes of the form $n^2 + 1$ (i.e., one more than a perfect square)?

- Are there infinitely many pairs of twin primes (i.e., primes that differ by 2, like 5 and 7 or 41 and 43)?

- Is Goldbach's conjecture true? (Euler failed to prove it; so has everyone since).

Although there has been no lack of effort, these questions remain open. Perhaps, like Fermat's last theorem, they will eventually be resolved. Or perhaps they will remain as challenges into the indefinite future. In order to spur research efforts across a wide range of mathematical disciplines, the privately funded Clay Mathematics Institute of Cambridge, Massachusetts, named seven "Millennium Prize Problems" in 2000, each with a million-dollar award for a correct solution. In any case, these mysteries justify Eric Temple Bell's characterization of number theory as "the last great uncivilized continent of mathematics."

The theory of numbers, then, is a vast and challenging subject as old as mathematics and as fresh as today's news. Its problems retain their fascination because of an apparent (often deceptive) simplicity and an irresistible beauty. With such a rich and colorful history, number theory surely deserves to be called, in the famous words of Gauss, "the queen of mathematics."

Addition

Addition is an operation that finds the total number when two or more numbers are put together. In other words, addition is the process to find the sum of two or more numbers.

Examples of Addition:

- 3 and 4 add up to give 7.

- 50 and 50 add up to give 100.

Examples:

$$31 + 45 + 71 + 2$$

Choices:

A. 149

B. 139

C. 147

D. 150

Correct Answer: A

Solution:

Step 1: $31 + 45 + 71 + 2 = 149$

Subtraction

Subtraction is an arithmetic operation that represents the operation of removing objects from a collection. The result of a subtraction is called a difference. Subtraction is signified by the minus sign (–). For example, in the adjacent picture, there are $5 - 2$ apples—meaning 5 apples with 2 taken away, which is a total of 3 apples. Therefore, the *difference* of 5 and 2 is 3, that is, $5 - 2 = 3$. Subtraction represents removing or decreasing physical and abstract quantities using different kinds of objects including negative numbers, fractions, irrational numbers, vectors, decimals, functions, and matrices.

Subtraction follows several important patterns. It is anticommutative, meaning that changing the order changes the sign of the answer. It is also not associative, meaning that when one subtracts more than two numbers, the order in which subtraction is performed matters. Because 0 is the additive identity, subtraction of it does not change a number. Subtraction also obeys predictable rules concerning related operations such as addition and multiplication. All of these rules can be proven, starting with the subtraction of integers and generalizing up through the real numbers and beyond. General binary operations that continue these patterns are studied in abstract algebra.

Performing subtraction is one of the simplest numerical tasks. Subtraction of very small numbers is accessible to young children. In primary education, students are taught to subtract numbers in the decimal system, starting with single digits and progressively tackling more difficult problems.

In advanced algebra and in computer algebra, an expression involving subtraction like $A - B$ is generally treated as a shorthand notation for the addition $A + (-B)$. Thus, $A - B$ contains two terms, namely A and $-B$. This allows an easier use of associativity and commutativity.

Notation and Terminology

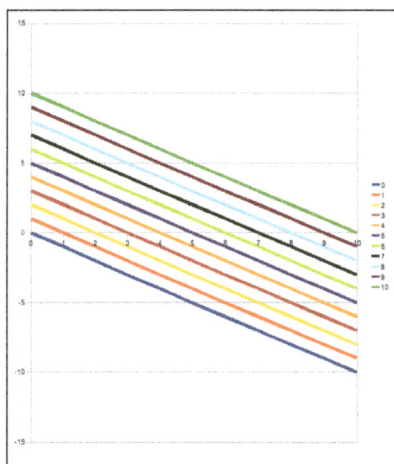

Subtraction of numbers 0–10. Line labels = minuend. X axis = subtrahend. Y axis = difference.

Subtraction is written using the minus sign "−" between the terms; that is, in infix notation. The result is expressed with an equals sign. For example,

$2 - 1 = 1$ (verbally, "two minus one equals one")

$4 - 2 = 2$ (verbally, "four minus two equals two")

$6 - 3 = 3$ (verbally, "six minus three equals three")

$4 - 6 = -2$ (verbally, "four minus six equals negative two")

There are also situations where subtraction is "understood" even though no symbol appears:

A column of two numbers, with the lower number in red, usually indicates that the lower number in the column is to be subtracted, with the difference written below, under a line. This is most common in accounting.

Formally, the number being subtracted is known as the *subtrahend*, while the number it is subtracted from is the *minuend*. The result is the *difference*.

All of this terminology derives from Latin. "Subtraction" is an English word derived from the Latin verb *subtrahere*, which is in turn a compound of *sub* "from under" and *trahere* "to pull"; thus to *subtract* is to *draw from below, take away*. Using the gerundive suffix *-nd* results in "subtrahend", "thing to be subtracted". Likewise from *minuere* "to reduce or diminish", one gets "minuend", "thing to be diminished".

Of Integers and Real Numbers

Integers

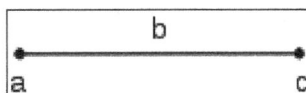

Imagine a line segment of length b with the left end labeled a and the right end labeled c. Starting from a, it takes b steps to the right to reach c. This movement to the right is modeled mathematically by addition:

$$a + b = c.$$

From c, it takes b steps to the left to get back to a. This movement to the left is modeled by subtraction:

$$c - b = a.$$

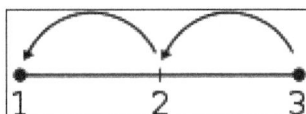

Now, a line segment labeled with the numbers 1, 2, and 3. From position 3, it takes no steps to the left to stay at 3, so $3 - 0 = 3$. It takes 2 steps to the left to get to position 1, so $3 - 2 = 1$. This picture is inadequate to describe what would happen after going 3 steps to the left of position 3. To represent such an operation, the line must be extended.

To subtract arbitrary natural numbers, one begins with a line containing every natural number $(0, 1, 2, 3, 4, 5, 6, ...)$. From 3, it takes 3 steps to the left to get to 0, so $3 - 3 = 0$. But $3 - 4$ is still invalid since it again leaves the line. The natural numbers are not a useful context for subtraction.

The solution is to consider the integer number line $(...,-3,-2,-1, 0, 1, 2, 3, ...)$. From 3, it takes 4 steps to the left to get to −1:

$$3 - 4 = -1.$$

Natural Numbers

Subtraction of natural numbers is not closed. The difference is not a natural number unless the minuend is greater than or equal to the subtrahend. For example, 26 cannot be subtracted from 11 to give a natural number. Such a case uses one of two approaches:

- Say that 26 cannot be subtracted from 11; subtraction becomes a partial function.
- Give the answer as an integer representing a negative number, so the result of subtracting 26 from 11 is −15.

Real Numbers

Subtraction of real numbers is defined as addition of signed numbers. Specifically, a number is subtracted by adding its additive inverse. Then we have $3 - \pi = 3 + (-\pi)$. This helps to keep the ring of real numbers "simple" by avoiding the introduction of "new" operators such as subtraction. Ordinarily a ring only has two operations defined on it; in the case of the integers, these are addition and multiplication. A ring already has the concept of additive inverses, but it does not have any notion of a separate subtraction operation, so the use of signed addition as subtraction allows us to apply the ring axioms to subtraction without needing to prove anything.

Properties

Anticommutativity

Subtraction is anti-commutative, meaning that if one reverses the terms in a difference left-to-right, the result is the negative of the original result. Symbolically, if a and b are any two numbers, then:

$$a - b = -(b - a).$$

Non-associativity

Subtraction is non-associative, which comes up when one tries to define repeated subtraction. Should the expression:

$$"a - b - c"$$

be defined to mean $(a - b) - c$ or $a - (b - c)$? These two possibilities give different answers. To resolve this issue, one must establish an order of operations, with different orders giving different results.

Predecessor

In the context of integers, subtraction of one also plays a special role: for any integer a, the integer $(a - 1)$ is the largest integer less than a, also known as the predecessor of a.

Units of Measurement

When subtracting two numbers with units of measurement such as kilograms or pounds, they must have the same unit. In most cases the difference will have the same unit as the original numbers.

Percentages

Changes in percentages can be reported in at least two forms, percentage change and percentage point change. Percentage change represents the relative change between the two quantities as a percentage, while percentage point change is simply the number obtained by subtracting the two percentages.

As an example, suppose that 30% of widgets made in a factory are defective. Six months later, 20% of widgets are defective. The percentage change is $-33\frac{1}{3}\%$, while the percentage point change is -10 percentage points.

In Computing

The method of complements is a technique used to subtract one number from another using only addition of positive numbers. This method was commonly used in mechanical calculators and is still used in modern computers.

Binary Digit	Ones' Complement
0	1
1	0

To subtract a binary number y (the subtrahend) from another number x (the minuend), the ones' complement of y is added to x and one is added to the sum. The leading digit "1" of the result is then discarded.

The method of complements is especially useful in binary (radix 2) since the ones' complement is very easily obtained by inverting each bit (changing "0" to "1" and vice versa). And adding 1 to get the two's complement can be done by simulating a carry into the least significant bit. For example:

    01100100   (x, equals decimal 100)

    − 00010110   (y, equals decimal 22)

becomes the sum:

      01100100   (x)

    + 11101001   (ones' complement of y)

    +   1   (to get the two's complement)
      _____
      101001110

Dropping the initial "1" gives the answer: 01001110 (equals decimal 78).

Teaching of Subtraction in Schools

Methods used to teach subtraction to elementary school vary from country to country, and within a country, different methods are in fashion at different times. In what is, in the United States, called traditional mathematics, a specific process is taught to students at the end of the 1st year or during the 2nd year for use with multi-digit whole numbers, and is extended in either the fourth or fifth grade to include decimal representations of fractional numbers.

In America

Almost all American schools currently teach a method of subtraction using borrowing or regrouping (the decomposition algorithm) and a system of markings called crutches. Although a method of borrowing had been known and published in textbooks previously, the use of crutches in American schools spread after William A. Brownell published a study claiming that crutches were beneficial to students using this method. This system caught on rapidly, displacing the other methods of subtraction in use in America at that time.

In Europe

Some European schools employ a method of subtraction called the Austrian method, also known as the additions method. There is no borrowing in this method. There are also crutches (markings to aid memory), which vary by country.

Comparing the two Main Methods

Both these methods break up the subtraction as a process of one digit subtractions by place value. Starting with a least significant digit, a subtraction of subtrahend:

$$s_j s_{j-1} \ldots s_1$$

from minuend:

$$m_k m_{k-1} \ldots m_1,$$

where each s_i and m_i is a digit, proceeds by writing down $m_1 - s_1, m_2 - s_2,$ and so forth, as long as s_i s_i does not exceed m_i. Otherwise, m_i is increased by 10 and some other digit is modified to correct for this increase. The American method corrects by attempting to decrease the minuend digit m_{i+1} by one (or continuing the borrow leftwards until there is a non-zero digit from which to borrow). The European method corrects by increasing the subtrahend digit s_{i+1} by one.

Example: 704 − 512 = 192.

−1			← carry
C	D	U	
7	0	4	← Minuend
5	1	2	← Subtrahend
1	9	2	← Rest or Difference

The minuend is 704, the subtrahend is 512. The minuend digits are $m_3 = 7, m_2 = 0$ and $m_1 = 4$. The subtrahend digits are $s_3 = 5, s_2 = 1$ and $s_1 = 2$. Beginning at the one's place, 4 is not less than 2 so the difference 2 is written down in the result's one's place. In the ten's place, 0 is less than 1, so the 0 is increased by 10, and the difference with 1, which is 9, is written down in the ten's place. The American method corrects for the increase of ten by reducing the digit in the minuend's hundreds place by one. That is, the 7 is struck through and replaced by a 6. The subtraction then proceeds in the hundreds place, where 6 is not less than 5, so the difference is written down in the result's hundred's place. We are now done, the result is 192.

The Austrian method does not reduce the 7 to 6. Rather it increases the subtrahend hundred's digit by one. A small mark is made near or below this digit (depending on the school). Then the subtraction proceeds by asking what number when increased by 1, and 5 is added to it, makes 7. The answer is 1, and is written down in the result's hundred's place.

There is an additional subtlety in that the student always employs a mental subtraction table in the American method. The Austrian method often encourages the student to mentally use the addition table in reverse. In the example above, rather than adding 1 to 5, getting 6, and subtracting that from 7, the student is asked to consider what number, when increased by 1, and 5 is added to it, makes 7.

Subtraction by Hand

Austrian Method

Example:

753 -491 ──── 2	753 -491 ──── 2	753 -491 ──── 2	753 -491 1 ──── 2
1 + … = 3	The difference is written under the line.	9 + … = 5 The required sum (5) is too small.	So, we add 10 to it and put a 1 under the next higher place in the subtrahend.
753 -491 1 ──── 62	753 -491 1 ──── 62	753 -491 1 ──── 262	753 -491 1 ──── 262
9 + … = 15 Now we can find the difference like before.	(4 + 1) + … = 7	The difference is written under the line.	The total difference.

Subtraction from Left to Right

Example:

753 -491 ──── 3	753 -491 ──── 2	753 -491 ──── 26	753 -491 ──── 26	753 -491 ──── 262
7 − 4 = 3 This result is only penciled in.	Because the next digit of the minuend is smaller than the subtrahend, we subtract one from our penciled-in-number and mentally add ten to the next.	15 − 9 = 6	Because the next digit in the minuend is not smaller than the subtrahend, We keep this number.	3 − 1 = 2

American Method

In this method, each digit of the subtrahend is subtracted from the digit above it starting from right to left. If the top number is too small to subtract the bottom number from it, we add 10 to it; this 10 is "borrowed" from the top digit to the left, which we subtract 1 from. Then we move on to subtracting the next digit and borrowing as needed, until every digit has been subtracted. Example:

$$
\begin{array}{r} 753 \\ -491 \\ \hline \end{array}
$$

3 – 1 = ...

$$
\begin{array}{r} 753 \\ -491 \\ \hline 2 \end{array}
$$

We write the difference under the line.

$$
\begin{array}{r} 753 \\ -491 \\ \hline 2 \end{array}
$$

5 – 9 = ...
The minuend (5) is too small.

$$
\begin{array}{r} {}^{6\ 15}753 \\ -491 \\ \hline 2 \end{array}
$$

So, we add 10 to it. The 10 is "borrowed" from the digit on the left, which goes down by 1.

$$
\begin{array}{r} {}^{6\ 15}753 \\ -491 \\ \hline 62 \end{array}
$$

15 – 9 = ...
Now the subtraction works, and we write the difference under the line.

$$
\begin{array}{r} {}^{6\ 15}753 \\ -491 \\ \hline 62 \end{array}
$$

6 – 4 = ...

$$
\begin{array}{r} {}^{6\ 15}753 \\ -491 \\ \hline 262 \end{array}
$$

We write the difference under the line.

$$
\begin{array}{r} {}^{6\ 15}753 \\ -491 \\ \hline 262 \end{array}
$$

The total difference.

Trade First

A variant of the American method where all borrowing is done before all subtraction.

Example:

$$
\begin{array}{r} {}^{4\ 11}751 \\ -493 \\ \hline \end{array}
$$

1 – 3 = not possible.
We add a 10 to the 1. Because the 10 is "borrowed" from the nearby 5, the 5 is lowered by 1.

$$
\begin{array}{r} {}^{6\ 14}_{4\ 11}751 \\ -493 \\ \hline \end{array}
$$

4 – 9 = not possible.
So we proceed as in step 1.

$$
\begin{array}{r} {}^{6\ 14}_{4\ 11}751 \\ -493 \\ \hline 8 \end{array}
$$

Working from right to left:
11 – 3 = 8

$$
\begin{array}{r} {}^{6\ 14}_{4\ 11}751 \\ -493 \\ \hline 58 \end{array}
$$

14 – 9 = 5

$$
\begin{array}{r} {}^{6\ 14}_{4\ 11}751 \\ -493 \\ \hline 258 \end{array}
$$

6 – 4 = 2

Partial Differences

The partial differences method is different from other vertical subtraction methods because no borrowing or carrying takes place. In their place, one places plus or minus signs depending on whether the minuend is greater or smaller than the subtrahend. The sum of the partial differences is the total difference.

Example:

753	753	753	753
-491	-491	-491	-491
+ 3 0 0	+ 3 0 0	+ 3 0 0	+ 3 0 0
		- 4 0	- 4 0
			+ 2

| The smaller number is subtracted from the greater: $700 - 400 = 300$ Because the minuend is greater than the subtrahend, this difference has a plus sign. | The smaller number is subtracted from the greater: $90 - 50 = 40$ Because the minuend is smaller than the subtrahend, this difference has a minus sign. | The smaller number is subtracted from the greater: $3 - 1 = 2$ Because the minuend is greater than the subtrahend, this difference has a plus sign. | $+300 - 40 + 2 = 262$ |

Nonvertical Methods

Counting Up

Instead of finding the difference digit by digit, one can count up the numbers between the subtrahend and the minuend.

Example: $1234 - 567 =$ can be found by the following steps:

- $567 + 3 = 570$
- $570 + 30 = 600$
- $600 + 400 = 1000$
- $1000 + 234 = 1234$

Add up the value from each step to get the total difference: $3 + 30 + 400 + 234 = 667$.

Breaking up the Subtraction

Another method that is useful for mental arithmetic is to split up the subtraction into small steps.

Example: $1234 - 567 =$ can be solved in the following way:

- $1234 - 500 = 734$
- $734 - 60 = 674$
- $674 - 7 = 667$

Same Change

The same change method uses the fact that adding or subtracting the same number from the minuend and subtrahend does not change the answer. One adds the amount needed to get zeros in the subtrahend.

Example:

1234 − 567 = can be solved as follows:

$$1234 - 567 = 1237 - 570 = 1267 - 600 = 667$$

Multiplication

Multiplication (often denoted by the cross symbol "×", by a point "·", by juxtaposition, or, on computers, by an asterisk "∗") is one of the four elementary mathematical operations of arithmetic, with the others being addition, subtraction and division.

The multiplication of whole numbers may be thought as a repeated addition; that is, the multiplication of two numbers is equivalent to adding as many copies of one of them, the *multiplicand*, as the value of the other one, the *multiplier*. The multiplier can be written first and multiplicand second (though the custom can vary by culture); both can be called *factors*.

$$a \times b = \underbrace{b + \cdots + b}_{a}$$

For example, 4 multiplied by 3 (often written as 3×4 and spoken as "3 times 4") can be calculated by adding 3 copies of 4 together:

$$3 \times 4 = 4 + 4 + 4 = 12$$

Here 3 and 4 are the *factors* and 12 is the *product*.

One of the main properties of multiplication is the commutative property: adding 3 copies of 4 gives the same result as adding 4 copies of 3:

$$4 \times 3 = 3 + 3 + 3 + 3 = 12$$

Thus the designation of multiplier and multiplicand does not affect the result of the multiplication.

The multiplication of integers (including negative numbers), rational numbers (fractions) and real numbers is defined by a systematic generalization of this basic definition.

Multiplication can also be visualized as counting objects arranged in a rectangle (for whole numbers) or as finding the area of a rectangle whose sides have given lengths. The area of a rectangle does not depend on which side is measured first, which illustrates the commutative property. The product of two measurements is a new type of measurement, for instance multiplying the lengths of the two sides of a rectangle gives its area, this is the subject of dimensional analysis.

The inverse operation of multiplication is division. For example, since 4 multiplied by 3 equals 12, then 12 divided by 3 equals 4. Multiplication by 3, followed by division by 3, yields the original number (since the division of a number other than 0 by itself equals 1).

Multiplication is also defined for other types of numbers, such as complex numbers, and more abstract constructs, like matrices. For some of these more abstract constructs, the order in which the operands are multiplied together matters.

Notation and Terminology

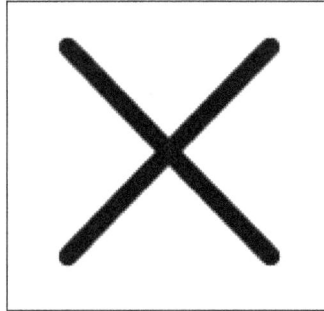

The multiplication sign ×.

In arithmetic, multiplication is often written using the sign "×" between the terms; that is, in infix notation. For example,

$2 \times 3 = 6$ (verbally, "two times three equals six")

$3 \times 4 = 12$

$2 \times 3 \times 5 = 6 \times 5 = 30$

$2 \times 2 \times 2 \times 2 \times 2 = 32$

There are other mathematical notations for multiplication:

- Multiplication is also denoted by dot signs, usually a middle-position dot (rarely period):

    $5 \cdot 2$ or $5 \cdot 3$

    The middle dot notation, encoded in Unicode as U+22C5 · DOT OPERATOR, is standard in the United States, the United Kingdom, and other countries where the period is used as a decimal point. When the dot operator character is not accessible, the interpunct (·) is used. In other countries that use a comma as a decimal mark, either the period or a middle dot is used for multiplication.

- In algebra, multiplication involving variables is often written as a juxtaposition (e.g., xy for x times y or $5x$ for five times x), also called *implied multiplication*. The notation can also be used for quantities that are surrounded by parentheses. This implicit usage of multiplication can cause ambiguity when the concatenated variables happen to match the name of another variable, when a variable name in front of a parenthesis can be confused with a function name, or in the correct determination of the order of operations.

- In vector multiplication, there is a distinction between the cross and the dot symbols. The cross symbol generally denotes the taking a cross product of two vectors, yielding a vector as the result, while the dot denotes taking the dot product of two vectors, resulting in a scalar.

In computer programming, the asterisk (as in 5*2) is still the most common notation. This is due to the fact that most computers historically were limited to small character sets (such as ASCII and EBCDIC) that lacked a multiplication sign (such as · or ×), while the asterisk appeared on every keyboard. This usage originated in the FORTRAN programming language.

The numbers to be multiplied are generally called the "factors". The number to be multiplied is the "multiplicand", and the number by which it is multiplied is the "multiplier". Usually the multiplier is placed first and the multiplicand is placed second; however sometimes the first factor is the multiplicand and the second the multiplier.

Also as the result of a multiplication does not depend on the order of the factors, the distinction between "multiplicand" and "multiplier" is useful only at a very elementary level and in some multiplication algorithms, such as the long multiplication. Therefore, in some sources, the term "multiplicand" is regarded as a synonym for "factor". In algebra, a number that is the multiplier of a variable or expression (e.g., the 3 in $3xy^2$) is called a coefficient.

The result of a multiplication is called a product. A product of integers is a multiple of each factor. For example, 15 is the product of 3 and 5, and is both a multiple of 3 and a multiple of 5.

Computation

The common methods for multiplying numbers using pencil and paper require a multiplication table of memorized or consulted products of small numbers (typically any two numbers from 0 to 9), however one method, the peasant multiplication algorithm, does not.

Multiplying numbers to more than a couple of decimal places by hand is tedious and error prone. Common logarithms were invented to simplify such calculations, since adding logarithms is equivalent to multiplying. The slide rule allowed numbers to be quickly multiplied to about three places of accuracy.

Beginning in the early 20th century, mechanical calculators, such as the Marchant, automated multiplication of up to 10 digit numbers. Modern electronic computers and calculators have greatly reduced the need for multiplication by hand.

Historical Algorithms

Methods of multiplication were documented in the Egyptian, Greek, Indian and Chinese civilizations. The Ishango bone, dated to about 18,000 to 20,000 BC, hints at a knowledge of multiplication in the Upper Paleolithic era in Central Africa.

Egyptians

The Egyptian method of multiplication of integers and fractions, documented in the Ahmes Papyrus, was by successive additions and doubling. For instance, to find the product of 13 and 21 one had to double 21 three times, obtaining 2 × 21 = 42, 4 × 21 = 2 × 42 = 84, 8 × 21 = 2 × 84 = 168. The full product could then be found by adding the appropriate terms found in the doubling sequence:

$$13 \times 21 = (1+4+8) \times 21 = (1 \times 21) + (4 \times 21) + (8 \times 21) = 21 + 84 + 168 = 273.$$

Babylonians

The Babylonians used a sexagesimal positional number system, analogous to the modern day decimal system. Thus, Babylonian multiplication was very similar to modern decimal multiplication. Because of the relative difficulty of remembering 60×60 different products, Babylonian mathematicians employed multiplication tables. These tables consisted of a list of the first twenty multiples of a certain *principal number* $n:n, 2n, ..., 20n$; followed by the multiples of $10n:30n, 40n$, and $50n$. Then to compute any sexagesimal product, say $53n$, one only needed to add $50n$ and $3n$ computed from the table.

Chinese

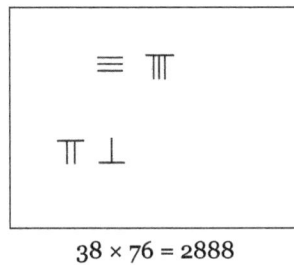

$38 \times 76 = 2888$

In the mathematical text *Zhoubi Suanjing*, dated prior to 300 BC, and the *Nine Chapters on the Mathematical Art*, multiplication calculations were written out in words, although the early Chinese mathematicians employed Rod calculus involving place value addition, subtraction, multiplication and division. Chinese were already using a decimal multiplication table since the Warring States period.

Modern Methods

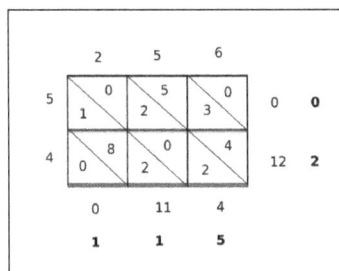

Product of 45 and 256. Note the order of the numerals in 45 is reversed down the left column. The carry step of the multiplication can be performed at the final stage of the calculation (in bold), returning the final product of $45 \times 256 = 11520$. This is a variant of Lattice multiplication.

The modern method of multiplication based on the Hindu–Arabic numeral system was first described by Brahmagupta. Brahmagupta gave rules for addition, subtraction, multiplication and division. Henry Burchard Fine, then professor of Mathematics at Princeton University, wrote the following:

> The Indians are the inventors not only of the positional decimal system itself, but of most of the processes involved in elementary reckoning with the system. Addition and subtraction they performed quite as they are performed nowadays; multiplication they effected in many ways, ours among them, but division they did cumbrously.

These place value decimal arithmetic algorithms were introduced to Arab countries by Al Khwarizmi in the early 9th century, and popularized in the Western world by Fibonacci in the 13th century.

Grid Method

Grid method multiplication or the box method, is used in primary schools in England and Wales & in some areas of the United States to help teach an understanding of how multiple digit multiplication works. An example of multiplying 34 by 13 would be to lay the numbers out in a grid like:

	30	4
10	300	40
3	90	12

and then add the entries.

Computer Algorithms

The classical method of multiplying two $n-$ digit numbers requires n^2 digit multiplications. Multiplication algorithms have been designed that reduce the computation time considerably when multiplying large numbers. Methods based on the discrete Fourier transform reduce the computational complexity to $O(n \log n \log \log n)$. Recently, the factor $\log \log n$ has been replaced by a function that increases much slower although it is still not constant (as it can be hoped).

In March, 2019, David Harvey and Joris van der Hoeven submitted an article presenting an integer multiplication algorithm with a claimed complexity of $O(n \log n)$.

Products of Measurements

One can only meaningfully add or subtract quantities of the same type but can multiply or divide quantities of different types. Four bags with three marbles each can be thought of as:

[4 bags] × [3 marbles per bag] = 12 marbles.

When two measurements are multiplied together the product is of a type depending on the types of the measurements. The general theory is given by dimensional analysis. This analysis is routinely applied in physics but has also found applications in finance.

A common example is multiplying speed by time gives distance, so:

50 kilometers per hour × 3 hours = 150 kilometers.

In this case, the hour units cancel out and we are left with only kilometer units.

Other examples:

2.5 meters × 4.5 meters = 11.25 square meters.

11 meters/seconds × 9 seconds = 99 meters.

4.5 residents per house × 20 houses = 90 residents.

Products of Sequences

Capital Pi Notation

The product of a sequence of terms can be written with the product symbol, which derives from the capital letter Π (Pi) in the Greek alphabet. Unicode position $U + 220F$ $(\prod)$ contains a glyph for denoting such a product, distinct from $U + 03A0$ (Π), the letter. The meaning of this notation is given by:

$$\prod_{i=1}^{4} i = 1 \cdot 2 \cdot 3 \cdot 4,$$

that is:

$$\prod_{i=1}^{4} i = 24.$$

The subscript gives the symbol for a dummy variable (i in this case), called the "index of multiplication" together with its lower bound (1), whereas the superscript (here 4) gives its upper bound. The lower and upper bound are expressions denoting integers. The factors of the product are obtained by taking the expression following the product operator, with successive integer values substituted for the index of multiplication, starting from the lower bound and incremented by 1 up to and including the upper bound. So, for example:

$$\prod_{i=1}^{6} i = 1 \cdot 2 \cdot 3 \cdot 4 \cdot 5 \cdot 6 = 720$$

More generally, the notation is defined as:

$$\prod_{i=m}^{n} x_i = x_m \cdot x_{m+1} \cdot x_{m+2} \cdots \cdots x_{n-1} \cdot x_n,$$

where m and n are integers or expressions that evaluate to integers. In case $m = n$, the value of the product is the same as that of the single factor x_m. If $m > n$, the product is the empty product, with the value 1.

Infinite Products

One may also consider products of infinitely many terms; these are called infinite products. Notationally, we would replace n above by the lemniscate ∞. The product of such a series is defined as the limit of the product of the first n terms, as n grows without bound. That is, by definition,

$$\prod_{i=m}^{\infty} x_i = \lim_{n \to \infty} \prod_{i=m}^{n} x_i.$$

One can similarly replace m with negative infinity, and define:

$$\prod_{i=-\infty}^{\infty} x_i = \left(\lim_{m \to -\infty} \prod_{i=m}^{0} x_i \right) \cdot \left(\lim_{n \to \infty} \prod_{i=1}^{n} x_i \right),$$

provided both limits exist.

Properties

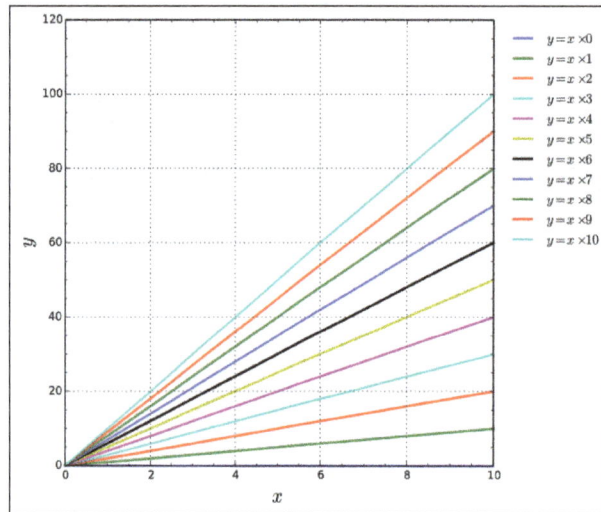

Multiplication of numbers 0–10. Line labels = multiplicand. X axis = multiplier. Y axis = product. Extension of this pattern into other quadrants gives the reason why a negative number times a negative number yields a positive number.

For the real and complex numbers, which includes for example natural numbers, integers, and fractions, multiplication has certain properties:

Commutative Property

The order in which two numbers are multiplied does not matter:

$$x \cdot y = y \cdot x.$$

Associative Property

Expressions solely involving multiplication or addition are invariant with respect to order of operations:

$$(x \cdot y) \cdot z = x \cdot (y \cdot z)$$

Distributive Property

Holds with respect to multiplication over addition. This identity is of prime importance in simplifying algebraic expressions:

$$x \cdot (y + z) = x \cdot y + x \cdot z$$

Identity Element

The multiplicative identity is 1; anything multiplied by 1 is itself. This feature of 1 is known as the identity property:

$$x \cdot 1 = x$$

Property of 0

Any number multiplied by 0 is 0. This is known as the zero property of multiplication:

$$x \cdot 0 = 0$$

Negation

−1 times any number is equal to the additive inverse of that number.

$$(-1) \cdot x = (-x) \text{ where } (-x) + x = 0$$

−1 times −1 is 1.

$$(-1) \cdot (-1) = 1$$

Inverse Element

Every number x, except 0, has a multiplicative inverse, $\dfrac{1}{x}$, such that $x \cdot \left(\dfrac{1}{x}\right) = 1$.

Order Preservation

Multiplication by a positive number preserves order:

For $a > 0$, if $b > c$ then $ab > ac$.

Multiplication by a negative number reverses order:

For $a < 0$, if $b > c$ then $ab < ac$.

The complex numbers do not have an ordering.

Other mathematical systems that include a multiplication operation may not have all these properties. For example, multiplication is not, in general, commutative for matrices and quaternions.

Axioms

In the book *Arithmetices principia, nova methodo exposita*, Giuseppe Peano proposed axioms for arithmetic based on his axioms for natural numbers. Peano arithmetic has two axioms for multiplication:

$$x \times 0 = 0$$

$$x \times S(y) = (x \times y) + x$$

Here $S(y)$ represents the successor of y, or the natural number that *follows* y. The various properties like associativity can be proved from these and the other axioms of Peano arithmetic including induction. For instance $S(0)$, denoted by 1, is a multiplicative identity because:

$$x \times 1 = x \times S(0) = (x \times 0) + x = 0 + x = x$$

The axioms for integers typically define them as equivalence classes of ordered pairs of natural numbers. The model is based on treating (x,y) as equivalent to $x - y$ when x and y are treated as integers. Thus both $(0,1)$ and $(1,2)$ are equivalent to -1. The multiplication axiom for integers defined this way is:

$$(x_p, x_m) \times (y_p, y_m) = (x_p \times y_p + x_m \times y_m, x_p \times y_m + x_m \times y_p)$$

The rule that $-1 \times -1 = 1$ can then be deduced from:

$$(0,1) \times (0,1) = (0 \times 0 + 1 \times 1, 0 \times 1 + 1 \times 0) = (1,0)$$

Multiplication is extended in a similar way to rational numbers and then to real numbers.

Multiplication with Set Theory

The product of non-negative integers can be defined with set theory using cardinal numbers or the Peano axioms. How to extend this to multiplying arbitrary integers, and then arbitrary rational numbers. The product of real numbers is defined in terms of products of rational numbers.

Multiplication in Group Theory

There are many sets that, under the operation of multiplication, satisfy the axioms that define group structure. These axioms are closure, associativity, and the inclusion of an identity element and inverses.

A simple example is the set of non-zero rational numbers. Here we have identity 1, as opposed to groups under addition where the identity is typically 0. Note that with the rationals, we must exclude zero because, under multiplication, it does not have an inverse: there is no rational number that can be multiplied by zero to result in 1. In this example we have an abelian group, but that is not always the case.

Look at the set of invertible square matrices of a given dimension, over a given field. Now it is straightforward to verify closure, associativity, and inclusion of identity (the identity matrix) and inverses. However, matrix multiplication is not commutative, therefore this group is nonabelian.

Another fact of note is that the integers under multiplication is not a group, even if we exclude zero. This is easily seen by the nonexistence of an inverse for all elements other than 1 and -1.

Multiplication in group theory is typically notated either by a dot, or by juxtaposition (the omission of an operation symbol between elements). So multiplying element a by element b could be notated a . b or ab. When referring to a group via the indication of the set and operation, the dot is used, e.g., our first example could be indicated by $(\mathbb{Q} \setminus \{0\}, \cdot)$

Multiplication of Different Kinds of Numbers

Numbers can *count* (3 apples), *order* (the 3rd apple), or *measure* (3.5 feet high); as the history of mathematics has progressed from counting on our fingers to modelling quantum mechanics, multiplication has been generalized to more complicated and abstract types of numbers, and to things that are not numbers (such as matrices) or do not look much like numbers (such as quaternions).

Integers

$N \times M$ is the sum of N copies of M when N and M are positive whole numbers. This gives the number of things in an array N wide and M high.

Generalization to negative numbers can be done by:

$$N \times (-M) = (-N) \times M = -(N \times M) \text{ and } (-N) \times (-M) = N \times M$$

The same sign rules apply to rational and real numbers.

Rational Numbers

Generalization to fractions $\frac{A}{B} \times \frac{C}{D}$ is by multiplying the numerators and denominators respectively: $\frac{A}{B} \times \frac{C}{D} = \frac{(A \times C)}{(B \times D)}$. This gives the area of a rectangle $\frac{A}{B}$ high and $\frac{C}{D}$ wide, and is the same as the number of things in an array when the rational numbers happen to be whole numbers.

Real Numbers

Real numbers and their products can be defined in terms of sequences of rational numbers.

Complex Numbers

Considering complex numbers z_1 and z_2 as ordered pairs of real numbers (a_1, b_1) and (a_2, b_2), the product $z_1 \times z_2$ is $(a_1 \times a_2 - b_1 \times b_2, a_1 \times b_2 + a_2 \times b_1)$. This is the same as for reals, $a_1 \times a_2$, when the *imaginary parts* b_1 and b_2 are zero.

Equivalently, denoting $\sqrt{-1}$ as i, we have:

$$z_1 \times z_2 = (a_1 + b_1 i)(a_2 + b_2 i) = (a_1 \times a_2) + (a_1 \times b_2 i) + (b_1 \times a_2 i) + (b_1 \times b_2 i^2) = (a_1 a_2 - b_1 b_2) + (a_1 b_2 + b_1 a_2)i.$$

Further Generalizations

A very general, and abstract, concept of multiplication is as the "multiplicatively denoted" (second) binary operation in a ring. An example of a ring that is not any of the above number systems is a polynomial ring (you can add and multiply polynomials, but polynomials are not numbers in any usual sense).

Division

Often division, $\frac{x}{y}$, is the same as multiplication by an inverse, $x\left(\frac{1}{y}\right)$. Multiplication for some types of "numbers" may have corresponding division, without inverses; in an integral domain x

may have no inverse "$\frac{1}{x}$" but $\frac{x}{y}$ may be defined. In a division ring there are inverses, but $\frac{x}{y}$ may be ambiguous in non-commutative rings since $x\left(\frac{1}{y}\right)$ need not be the same as $\left(\frac{1}{y}\right)x$.

Exponentiation

When multiplication is repeated, the resulting operation is known as exponentiation. For instance, the product of three factors of two (2×2×2) is "two raised to the third power", and is denoted by 2^3, a two with a superscript three. In this example, the number two is the base, and three is the exponent. In general, the exponent (or superscript) indicates how many times the base appears in the expression, so that the expression:

$$a^n = \underbrace{a \times a \times \cdots \times a}_{n}$$

indicates that n copies of the base a are to be multiplied together. This notation can be used whenever multiplication is known to be power associative.

Division

Division is one of the four basic operations of arithmetic, the others being addition, subtraction, and multiplication. The mathematical symbols used for the division operator are the obelus (÷), the colon (:) and the slash (/).

At an elementary level the division of two natural numbers is – among other possible interpretations – the process of calculating the number of times one number is contained within another one. This number of times is not always an integer, and this led to two different concepts.

The division with remainder or Euclidean division of two natural numbers provides a *quotient*, which is the number of times the second one is contained in the first one, and a *remainder*, which is the part of the first number that remains, when in the course of computing the quotient, no further full chunk of the size of the second number can be allocated.

For a modification of this division to yield only one single result, the natural numbers must be extended to rational numbers or real numbers. In these enlarged number systems, division is the inverse operation to multiplication, that is $a = c \div b$ means $a \times b = c$, as long as b is not zero. If $b = 0$, then this is a division by zero, which is not defined.

Both forms of divisions appear in various algebraic structures. Those in which a Euclidean division (with remainder) is defined are called Euclidean domains and include polynomial rings in one indeterminate. Those in which a division (with a single result) by all nonzero elements is defined are called fields and division rings. In a ring the elements by which division is always possible are called the units (for example, 1 and −1 in the ring of integers).

In its simplest form, division can be viewed either as a quotition or a partition. In terms of quotition, $20 \div 5$ means the number of 5s that must be added to get 20. In terms of partition, $20 \div 5$ means the size of each of 5 parts into which a set of size 20 is divided. For example, 20 apples divide into four groups of five apples, meaning that *twenty divided by five is equal to four*. This is denoted as $20 / 5 = 4, 20 \div 5 = 4,$, or $\dfrac{20}{5} = 4$. Notationally, the *dividend* is divided by the *divisor* to get a *quotient*. In the example, 20 is the dividend, 5 is the divisor, and 4 is the quotient.

Unlike the other basic operations, when dividing natural numbers there is sometimes a remainder that will not go evenly into the dividend; for example, $10 \div 3$ leaves a remainder of 1, as 10 is not a multiple of 3. Sometimes this remainder is added to the quotient as a fractional part, so $10 \div 3$ is equal to $3\dfrac{1}{3}$ or 3.33..., but in the context of integer division, where numbers have no fractional part, the remainder is kept separately or discarded. When the remainder is kept as a fraction, it leads to a rational number. The set of all rational numbers is created by every possible division using integers. In modern mathematical terms, this is known as *extending the system*.

Unlike multiplication and addition, Division is not commutative, meaning that $a \div b$ is not always equal to $b \div a$. Division is also not, in general, associative, meaning that when dividing multiple times, the order of division can change the result. For example, $(20 \div 5) \div 2 = 2$, but $20 \div (5 \div 2) = 8$ (where the use of parentheses indicates that the operations inside parentheses are performed before the operations outside parentheses).

Division is, however, distributive, in the sense that $(a + b) \div c = (a \div c) + (b \div c)$ for every number. Specifically, division has the right-distributive property over addition and subtraction. That means:

$$\frac{a+b}{c} = (a+b) \div c = \frac{a}{c} + \frac{b}{c}$$

This is the same as multiplication: $(a + b) \times c = a \times c + b \times c$. However, division is *not* left-distributive:

$$\frac{a}{b+c} = a \div (b+c) = \left(\frac{b}{a} + \frac{c}{a}\right)^{-1} \neq \frac{a}{b} + \frac{a}{c}$$

Which is unlike the case in multiplication.

If there are multiple divisions in a row, the order of calculation traditionally goes from left to right, which is called left-associative:

$$a \div b \div c = (a \div b) \div c = a \div (b \times c) = a \times b^{-1} \times c^{-1}.$$

Notation

Division is often shown in algebra and science by placing the *dividend* over the *divisor* with a horizontal line, also called a fraction bar, between them. For example, "*a* divided by *b*" can written as:

$$\frac{a}{b}$$

which can also be read out loud as "*a* by *b*" or "*a* over *b*". A way to express division all on one line is to write the *dividend* (or numerator), then a slash, then the *divisor* (or denominator), as follows:

$$a / b$$

This is the usual way of specifying division in most computer programming languages, since it can easily be typed as a simple sequence of ASCII characters. Some mathematical software, such as MATLAB and GNU Octave, allows the operands to be written in the reverse order by using the backslash as the division operator:

$$b \backslash a$$

A typographical variation halfway between these two forms uses a solidus (fraction slash) but elevates the dividend, and lowers the divisor:

$$a / b$$

Any of these forms can be used to display a fraction. A fraction is a division expression where both dividend and divisor are integers (typically called the *numerator* and *denominator*), and there is no implication that the division must be evaluated further. A second way to show division is to use the obelus (or division sign), common in arithmetic, in this manner:

$$a \div b$$

This form is infrequent except in elementary arithmetic. ISO 80000-2-9.6 states it should not be used. The obelus is also used alone to represent the division operation itself, as for instance as a label on a key of a calculator. The obelus was introduced by Swiss mathematician Johann Rahn in 1659 in *Teutsche Algebra*.

$$a : b$$

In some non-English-speaking countries colon is used to denote division. This notation was introduced by Gottfried Wilhelm Leibniz in his 1684 *Acta eruditorum*. Leibniz disliked having separate symbols for ratio and division. However, in English usage the colon is restricted to expressing the related concept of ratios.

Since the 19th century US textbooks have used $b)a$ or $b\overline{)a}$ to denote *a* divided by *b*, especially when discussing long division. The history of this notation is not entirely clear because it evolved over time.

Computing

Manual Methods

Division is often introduced through the notion of "sharing out" a set of objects, for example a pile of lollies, into a number of equal portions. Distributing the objects several at a time in each round of sharing to each portion leads to the idea of "chunking" — a form of division where one repeatedly subtracts multiples of the divisor from the dividend itself.

By allowing one to subtract more multiples than what the partial remainder allows at a given stage, more flexible methods, such as the bidirectional variant of chunking, can be developed as well.

More systematic and more efficient (but also more formalised, more rule-based, and more removed from an overall holistic picture of what division is achieving), a person who knows the multiplication tables can divide two integers with pencil and paper using the method of short division, if the divisor is small, or long division, if the divisor is larger.

If the dividend has a fractional part (expressed as a decimal fraction), one can continue the algorithm past the ones place as far as desired. If the divisor has a fractional part, one can restate the problem by moving the decimal to the right in both numbers until the divisor has no fraction.

A person can calculate division with an abacus by repeatedly placing the dividend on the abacus, and then subtracting the divisor the offset of each digit in the result, counting the number of divisions possible at each offset.

A person can use logarithm tables to divide two numbers, by subtracting the two numbers' logarithms, then looking up the antilogarithm of the result.

A person can calculate division with a slide rule by aligning the divisor on the C scale with the dividend on the D scale. The quotient can be found on the D scale where it is aligned with the left index on the C scale. The user is responsible, however, for mentally keeping track of the decimal point.

By Computer or with Computer Assistance

Modern computers compute division by methods that are faster than long division, with the more efficient ones relying on approximation techniques from numerical analysis.

In modular arithmetic (modulo a prime number) and for real numbers, nonzero numbers have a multiplicative inverse. In these cases, a division by x may be computed as the product by the multiplicative inverse of x. This approach is often associated with the faster methods in computer arithmetic.

Division in Different Contexts

Euclidean Division

The Euclidean division is the mathematical formulation of the outcome of the usual process of division of integers. It asserts that, given two integers, a, the *dividend*, and b, the *divisor*, such that

$b \neq 0$, there are unique integers q, the quotient, and r, the remainder, such that $a = bq + r$ and $0 \leq r < |b|$, where $|b|$ denotes the absolute value of b.

Integers

Integers are not closed under division. Apart from division by zero being undefined, the quotient is not an integer unless the dividend is an integer multiple of the divisor. For example, 26 cannot be divided by 11 to give an integer. Such a case uses one of five approaches:

- Say that 26 cannot be divided by 11; division becomes a partial function.

- Give an approximate answer as a decimal fraction or a mixed number, so $\frac{26}{11} \simeq 2.36$ or $\frac{26}{11} \simeq 2\frac{36}{100}$. This is the approach usually taken in numerical computation.
  Give the answer as a fraction representing a rational number, so the result of the division of 26 by 11 is $\frac{26}{11}$. But, usually, the resulting fraction should be simplified: the result of the division of 52 by 22 is also $\frac{26}{11}$. This simplification may be done by factoring out the greatest common divisor.

- Give the answer as an integer *quotient* and a *remainder*, so $\frac{26}{11} = 2$ remainder.

- To make the distinction with the previous case, this division, with two integers as result, is sometimes called *Euclidean division*, because it is the basis of the Euclidean algorithm.

- Give the integer quotient as the answer, so $\frac{26}{11} = 2$. This is sometimes called *integer division*.

Dividing integers in a computer program requires special care. Some programming languages, such as C, treat integer division as in case above, so the answer is an integer. Other languages, such as MATLAB and every computer algebra system return a rational number as the answer, as in case above. These languages also provide functions to get the results of the other cases, either directly or from the result of case.

Names and symbols used for integer division include div, /, \, and %. Definitions vary regarding integer division when the dividend or the divisor is negative: rounding may be toward zero (so called T-division) or toward $-\infty$ (F-division); rarer styles can occur.

Divisibility rules can sometimes be used to quickly determine whether one integer divides exactly into another.

Rational Numbers

The result of dividing two rational numbers is another rational number when the divisor is not 0. The division of two rational numbers p/q and r/s can be computed as:

$$\frac{p/q}{r/s} = \frac{p}{q} \times \frac{s}{r} = \frac{ps}{qr}.$$

All four quantities are integers, and only p may be 0. This definition ensures that division is the inverse operation of multiplication.

Real Numbers

Division of two real numbers results in another real number (when the divisor is nonzero). It is defined such that $a/b = c$ if and only if $a = cb$ and $b \neq 0$.

Complex Numbers

Dividing two complex numbers (when the divisor is nonzero) results in another complex number, which is found using the conjugate of the denominator:

$$\frac{p+iq}{r+is} = \frac{(p+iq)(r-is)}{(r+is)(r-is)} = \frac{pr+qs+i(qr-ps)}{r^2+s^2} = \frac{pr+qs}{r^2+s^2} + i\frac{qr-ps}{r^2+s^2}.$$

This process of multiplying and dividing by $r - is$ is called 'realisation' or (by analogy) rationalisation. All four quantities p, q, r, s are real numbers, and r and s may not both be 0.

Division for complex numbers expressed in polar form is simpler than the definition above:

$$\frac{pe^{iq}}{re^{is}} = \frac{pe^{iq}e^{-is}}{re^{is}e^{-is}} = \frac{p}{r}e^{i(q-s)}.$$

Again all four quantities p, q, r, s are real numbers, and r may not be 0.

Polynomials

One can define the division operation for polynomials in one variable over a field. Then, as in the case of integers, one has a remainder.

Matrices

One can define a division operation for matrices. The usual way to do this is to define $A/B = AB^{-1}$, where B^{-1} denotes the inverse of B, but it is far more common to write out AB^{-1} explicitly to avoid confusion. An elementwise division can also be defined in terms of the Hadamard product.

Left and Right Division

Because matrix multiplication is not commutative, one can also define a left division or so-called backslash-division as $A \backslash B = A^{-1}B$. For this to be well defined, B^{-1} need not exist, however A^{-1} does need to exist. To avoid confusion, division as defined by:

$$A/B = AB^{-1}$$

Is sometimes called right division or slash-division in this context.

With left and right division defined this way, $A/(BC)$ is in general not the same as $(A/B)/C$, nor is $(AB)\backslash C$ the same as $A\backslash(B\backslash C)$. However, $A/(BC) = (A/C)/B$ and $(AB)\backslash C = B\backslash(A\backslash C)$.

Pseudoinverse

To avoid problems when A^{-1} and/or B^{-1} do not exist, division can also be defined as multiplication by the pseudoinverse. That is, $A/B = AB^+$ and $A \backslash B = A^+B$, where A^+ and B^+ denote the pseudo-inverses of A and B.

Abstract Algebra

In abstract algebra, given a magma with binary operation * (which could nominally be termed multiplication), left division of b by a (written $a \backslash b$) is typically defined as the solution x to the equation $a * x = b$, if this exists and is unique. Similarly, right division of b by a (written b / a) is the solution y to the equation $y * a = b$. Division in this sense does not require * to have any particular properties (such as commutativity, associativity, or an identity element).

"Division" in the sense of "cancellation" can be done in any magma by an element with the cancellation property. Examples include matrix algebras and quaternion algebras. A quasigroup is a structure in which division is always possible, even without an identity element and hence inverses. In an integral domain, where not every element need have an inverse, *division* by a cancellative element a can still be performed on elements of the form ab or ca by left or right cancellation, respectively. If a ring is finite and every nonzero element is cancellative, then by an application of the pigeonhole principle, every nonzero element of the ring is invertible, and *division* by any nonzero element is possible. To learn about when *algebras* (in the technical sense) have a division operation, refer to the page on division algebras. In particular Bott periodicity can be used to show that any real normed division algebra must be isomorphic to either the real numbers R, the complex numbers C, the quaternions H, or the octonions O.

Calculus

The derivative of the quotient of two functions is given by the quotient rule:

$$\left(\frac{f}{g}\right)' = \frac{f'g - fg'}{g^2}.$$

Division by Zero

Division of any number by zero in most mathematical systems is undefined, because zero multiplied by any finite number always results in a product of zero. Entry of such an expression into most calculators produces an error message. However, in certain higher level mathematics division by zero is possible by the zero ring and algebras such as wheels. In these algebras, the meaning of division is different from traditional definitions.

Decimal

The decimal numeral system (also called base-ten positional numeral system, and occasionally called denary or decanary) is the standard system for denoting integer and non-integer numbers.

It is the extension to non-integer numbers of the Hindu–Arabic numeral system. The way of denoting numbers in the decimal system is often referred to as *decimal notation.*

A *decimal numeral*, or just *decimal*, or casually *decimal number*, refers generally to the notation of a number in the decimal numeral system. Decimals may sometimes be identified for containing a decimal separator (for example the "." in 10.00 or 3.14159). "Decimal" may also refer specifically to the digits after the decimal separator, such as in "3.14 is the approximation of π to *two decimals*".

The numbers that may be represented in the decimal system are the decimal fractions, that is the fractions of the form $a/10^n$, where a is an integer, and n is a non-negative integer.

The decimal system has been extended to *infinite decimals*, for representing any real number, by using an infinite sequence of digits after the decimal separator. In this context, the decimal numerals with a finite number of non–zero places after the decimal separator are sometimes called *terminating decimals*. A repeating decimal is an infinite decimal that after some place repeats indefinitely the same sequence of digits (for example $5.123144144144144... = 5.123\overline{144}$). An infinite decimal represents a rational number if and only if it is a repeating decimal or has a finite number of nonzero digits.

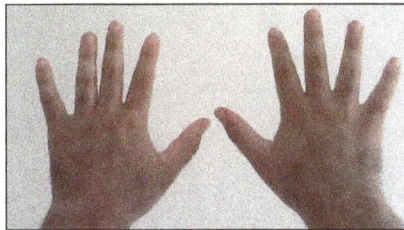

Ten fingers on two hands, the possible starting point of the decimal counting.

Many numeral systems of ancient civilisations use ten and its powers for representing numbers, possibly because there are ten fingers on two hands and people started counting by using their fingers. Examples are Brahmi numerals, Greek numerals, Hebrew numerals, Roman numerals, and Chinese numerals. Very large numbers were difficult to represent in these old numeral systems, and only the best mathematicians were able to multiply or divide large numbers. These difficulties were completely solved with the introduction of the Hindu–Arabic numeral system for representing integers. This system has been extended to represent some non-integer numbers, called *decimal fractions* or *decimal numbers* for forming the *decimal numeral system.*

Decimal Notation

For writing numbers, the decimal system uses ten decimal digits, a decimal mark, and, for negative numbers, a minus sign "–". The decimal digits are 0, 1, 2, 3, 4, 5, 6, 7, 8, 9; the decimal separator is the dot "." in many countries (including all English speaking ones), but may be a comma "," in other countries (mainly in continental Europe).

For representing a non-negative number, a decimal consists of

- Either a (finite) sequence of digits such as 2017, or in full generality,

  $$a_m a_{m-1} \ldots a_0$$

  (in this case, the (entire) decimal represents an integer).

- Two sequence of digits separated by a decimal mark such as 3.14159, 15.00, or in full generality:

$$a_m a_{m-1} \ldots a_0 . b_1 b_2 \ldots b_n$$

It is generally assumed that, if $m > 0$, the first digit a_m is not zero, but, in some circumstances, it may be useful to have one or more 0's on the left. This does not change the value represented by the decimal. For example, 3.14 = 03.14 = 003.14. Similarly, if b_n =0, it may be removed, and conversely, trailing zeros may be added without changing the represented number: for example, 15 = 15.0 = 15.00 and 5.2 = 5.20 = 5.200 . Sometimes the extra zeros are used for indicating the accuracy of a measurement. For example, 15.00 m may indicate that the measurement error is less than one centimeter (0.01 m), while 15 m may mean that the length is roughly fifteen meters, and that the error may exceed 10 cm.

For representing a negative number, a minus sign is placed before a_m.

The numeral $a_m a_{m-1} \ldots a_0 . b_1 b_2 \ldots b_n$ represents the number:

$$a_m 10^m + a_{m-1} 10^{m-1} + \cdots + a_0 10^0 + \frac{b_1}{10^1} + \frac{b_2}{10^2} + \cdots + \frac{b_n}{10^n}$$

Therefore, the contribution of each digit to the value of a number depends on its position in the numeral. That is, the decimal system is a positional numeral system.

Decimal Fractions

The numbers that are represented by decimal numerals are the decimal fractions (sometimes called decimal numbers), that is, the rational numbers that may be expressed as a fraction, the denominator of which is a power of ten. For example, the numerals 0.8, 14.89, 0.00024 represent the fractions $\frac{8}{10}, \frac{1489}{100}, \frac{24}{100000}$. More generally, a decimal with n digits after the separator represents the fraction with denominator 10^n, whose numerator is the integer obtained by removing the separator.

Expressed as a fully reduced fraction, the decimal numbers are those whose denominator is a product of a power of 2 and a power of 5. Thus the smallest denominators of decimal numbers are:

$$1 = 2^0 \cdot 5^0, 2 = 2^1 \cdot 5^0, 4 = 2^2 \cdot 5^0, 5 = 2^0 \cdot 5^1, 8 = 2^3 \cdot 5^0, 10 = 2^1 \cdot 5^1, 16 = 2^4 \cdot 5^0, 25 = 2^0 \cdot 5^2, \ldots$$

The integer part or integral part of a decimal is the integer written to the left of the decimal separator. For a non-negative decimal, it is the largest integer that is not greater than the decimal. The part from the decimal separator to the right is the fractional part, which equals the difference between the numeral and its integer part.

When the integral part of a numeral is zero, it may occur, typically in computing, that the integer part is not written (for example .1234, instead of 0.1234). In normal writing, this is generally avoided because of the risk of confusion between the decimal mark and other punctuation.

Real Number Approximation

Decimal numerals do not allow an exact representation for all real numbers, e.g. for the real number π. Nevertheless, they allow approximating every real number with any desired accuracy, e.g., the decimal 3.14159 approximates the real π, being less than 10^{-5} off; and so decimals are widely used in science, engineering and everyday life.

More precisely, for every real number x, and every positive integer n, there are two decimals L and u, with at most n digits after the decimal mark, such that $L \le x \le u$ and $(u - L) = 10^{-n}$.

Numbers are very often obtained as the result of a measurement. As measurements are generally afflicted with some measurement error with a known upper bound, the result of a measurement is well represented by a decimal with n digits after the decimal mark, as soon as the absolute measurement error is bounded from above by 10^{-n}. In practice, measurement results are often given with a certain number of digits after the decimal point, which indicate the error bounds. For example, although 0.080 and 0.08 denote the same decimal number, the numeral 0.080 suggests a measurement with an error less than 0.001, while the numeral 0.08 indicates an absolute error bounded by 0.01. In both cases, the true value of the measured quantity could be, for example, 0.0803 or 0.0796.

Infinite Decimal Expansion

For a real number x and an integer $n \ge 0$, let $[x]_n$ denote the (finite) decimal expansion of the greatest number that is not greater than x, which has exactly n digits after the decimal mark. Let d_i denote the last digit of $[x]_i$. It is straightforward to see that $[x]_n$ may be obtained by appending d_n to the right of $[x]_{n-1}$. This way one has:

$$[x]_n = [x]_0 . d_1 d_2 ... d_{n-1} d_n,$$

and the difference of $[x]_{n-1}$ and $[x]_n$ amounts to:

$$\left|[x]_n - [x]_{n-1}\right| = d_n \cdot 10^{-n} < 10^{-n+1},$$

which is either 0, if $d_n = 0$, or gets arbitrarily small, when n tends to infinity. According to the definition of a limit, x is the limit of $[x]_n$ when n tends to infinity. This is written as $x = \lim_{n \to \infty}[x]_n$ or

$$x = [x]_0 . d_1 d_2 ... d_n ...,$$

which is called an infinite decimal expansion of x.

Conversely, for any integer $[x]_0$ and any sequence of digits $(d_n)_{n=1}^{\infty}$ the (infinite) expression $[x]_0 . d_1 d_2 ... d_n ...$ is an *infinite decimal expansion* of a real number x. This expansion is unique if neither all d_n are equal to 9 nor all d_n are equal to 0 for n large enough (for all n greater than some natural number N).

If all d_N for $n > N$ equal to 9 and $[x]_n = [x]_0 . d_1 d_2 ... d_n$, the limit of the sequence $([x]_n)_{n=1}^{\infty}$ is the decimal fraction obtained by replacing the last digit that is not a 9, i.e.: d_N, by $d_N + 1$, and replacing all subsequent 9s by 0s.

Any such decimal fraction, i.e., $d_n = 0$ for $n > N$, may be converted to its equivalent infinite decimal expansion by replacing d_N by $d_N - 1$, and replacing all subsequent 0s by 9s.

In summary, every real number that is not a decimal fraction has a unique infinite decimal expansion. Each decimal fraction has exactly two infinite decimal expansions, one containing only 0s after some place, which is obtained by the above definition of $[x]_n$, and the other containing only 9s after some place, which is obtained by defining $[x]_n$ as the greatest number that is *less* than x, having exactly n digits after the decimal mark.

Rational Numbers

Long division allows computing the infinite decimal expansion of a rational number. If the rational number is a decimal fraction, the division stops eventually, producing a decimal numeral, which may be prolongated into an infinite expansion by adding infinitely many zeros. If the rational number is not a decimal fraction, the division may continue indefinitely. However, as all successive remainders are less than the divisor, there are only a finite number of possible remainders, and after some place, the same sequence of digits must be repeated indefinitely in the quotient. That is, one has a repeating decimal. For example,

$$\frac{1}{81} = 0.012345679012... \text{ (with the group 012345679 indefinitely repeating).}$$

Conversely, every eventually repeating sequence of digits is the infinite decimal expansion of a rational number. This is a consequence of the fact that the recurring part of a decimal representation is, in fact, an infinite geometric series which will sum to a rational number. For example,

$$0.0123123123... = \frac{123}{10000}\sum_{k=0}^{\infty}0.001^k = \frac{123}{10000}\frac{1}{1-0.001} = \frac{123}{9990} = \frac{41}{3330}$$

Decimal Computation

1/2	1	2	3	4	5	6	7	8	9	10	20	30	40	50	60	70	80	90	
45	90	180	270	360	450	540	630	720	810	900	1800	2700	3600	4500	5400	6300	7200	8100	90
40	80	160	240	320	400	480	560	640	720	800	1600	2400	3200	4000	4800	5600	6400	7200	80
35	70	140	210	280	350	420	490	560	630	700	1400	2100	2800	3500	4200	4900	5600	6300	70
30	60	120	180	240	300	360	420	480	540	600	1200	1800	2400	3000	3600	4200	4800	5400	60
25	50	100	150	200	250	300	350	400	450	500	1000	1500	2000	2500	3000	3500	4000	4500	50
20	40	80	120	160	200	240	280	320	360	400	800	1200	1600	2000	2400	2800	3200	3600	40
15	30	60	90	120	150	180	210	240	270	300	600	900	1200	1500	1800	2100	2400	2700	30
10	20	40	60	80	100	120	140	160	180	200	400	600	800	1000	1200	1400	1600	1800	20
5	10	20	30	40	50	60	70	80	90	100	200	300	400	500	600	700	800	900	10
4.5	9	18	27	36	45	54	63	72	81	90	180	270	360	450	540	630	720	810	9
4	8	16	24	32	40	48	56	64	72	80	160	240	320	400	480	560	640	720	8
3.5	7	14	21	28	35	42	49	56	63	70	140	210	280	350	420	490	560	630	7
3	6	12	18	24	30	36	42	48	54	60	120	180	240	300	360	420	480	540	6
2.5	5	10	15	20	25	30	35	40	45	50	100	150	200	250	300	350	400	450	5
2	4	8	12	16	20	24	28	32	36	40	80	120	160	200	240	280	320	360	4
1.5	3	6	9	12	15	18	21	24	27	30	60	90	120	150	180	210	240	270	3
1	2	4	6	8	10	12	14	16	18	20	40	60	80	100	120	140	160	180	2
0.5	1	2	3	4	5	6	7	8	9	10	20	30	40	50	60	70	80	90	1
0.25	0.5	1	1.5	2	2.5	3	3.5	4	4.5	5	10	15	20	25	30	35	40	45	1/2

Diagram of the world's earliest multiplication table from the Warring States period.

Most modern computer hardware and software systems commonly use a binary representation internally (although many early computers, such as the ENIAC or the IBM 650, used decimal representation internally). For external use by computer specialists, this binary representation is sometimes presented in the related octal or hexadecimal systems.

For most purposes, however, binary values are converted to or from the equivalent decimal values for presentation to or input from humans; computer programs express literals in decimal by default. (123.1, for example, is written as such in a computer program, even though many computer languages are unable to encode that number precisely).

Both computer hardware and software also use internal representations which are effectively decimal for storing decimal values and doing arithmetic. Often this arithmetic is done on data which are encoded using some variant of binary-coded decimal, especially in database implementations, but there are other decimal representations in use (such as in the new IEEE 754 Standard for Floating-Point Arithmetic).

Decimal arithmetic is used in computers so that decimal fractional results of adding (or subtracting) values with a fixed length of their fractional part always are computed to this same length of precision. This is especially important for financial calculations, e.g., requiring in their results integer multiples of the smallest currency unit for book keeping purposes. This is not possible in binary, because the negative powers of 10 have no finite binary fractional representation; and is generally impossible for multiplication (or division).

References

- Thomas Sonnabend (2010). Mathematics for Teachers: An Interactive Approach for Grades K–8. Brooks/Cole, Cengage Learning (Charles Van Wagner). P. 126. ISBN 978-0-495-56166-8.^ Smith, Dav

- Arithmetic, science: britannica.com, Retrieved 10 March, 2019

- "The Definitive Higher Math Guide to Long Division and Its Variants — for Integers". Math Vault. 2019-02-24. Retrieved 2019-06-24

- Euclid, number-theory, science: britannica.com, Retrieved 11 April, 2019

- Coppa, A.; et al. (6 April 2006), "Early Neolithic tradition of dentistry: Flint tips were surprisingly effective for drilling tooth enamel in a prehistoric population" (PDF), Nature, 440 (7085): 755–6, Bibcode:2006Natur.440..755C, doi:10.1038/440755a, PMID 16598247

- Derbyshire, John (2004). Prime Obsession: Bernhard Riemann and the Greatest Unsolved Problem in Mathematics. New York City: Penguin Books. ISBN 978-0-452-28525-5

- Eporfolio-template, students: westminstercollege.edu, Retrieved 12 May, 2019

- "Ancient bamboo slips for calculation enter world records book". The Institute of Archaeology, Chinese Academy of Social Sciences. Retrieved 10 May 2017

Geometry   3

- **Euclidean Geometry**

- **Analytic Geometry**

- **Non-euclidean Geometry**

- **Differential Geometry**

- **Projective Geometry**

Geometry is a sub-discipline of mathematics which deals with the study of points, lines, surfaces, shapes, size, relative position of figures, etc. Euclidean geometry, analytical geometry, non-Euclidean geometry, projective geometry, etc. are some of the branches of geometry. The topics elaborated in this chapter will help in gaining a better perspective of geometry.

Geometry is an original field of mathematics, and is indeed the oldest of all sciences, going back at least to the times of Euclid, Pythagoras, and other "natural philosophers" of ancient Greece. Initially, geometry was studied to understand the physical world we live in, and the tradition continues to this day. Witness for example, the spectacular success of Einstein's theory of general relativity, a purely geometric theory that describes gravitation in terms of the curvature of a four-dimensional "spacetime". However, geometry transcends far beyond physical applications, and it is not unreasonable to say that geometric ideas and methods have always permeated every field of mathematics.

In modern language, the central object of study in geometry is a manifold, which is an object that may have a complicated overall shape, but such that on small scales it "looks like" ordinary space of a certain dimension. For example, a 1-dimensional manifold is an object such that small pieces of it look like a line, although in general it looks like a curve rather than a straight line. A 2-dimensional manifold, on small scales, looks like a (curved) piece of paper – there are two independent directions in which we can move at any point. For example, the surface of the Earth is a 2-dimensional manifold. An n-dimensional manifold likewise looks locally like an ordinary n-dimensional space. This does not necessarily correspond to any notion of "physical space". As an example, the data of the position and velocity of N particles in a room is described by 6N independent variables, because each particle needs 3 numbers to describe its position and 3 more numbers to describe its velocity. Hence, the "configuration space" of this system is a 6N-dimensional manifold. If for some reason the motion of these particles were not independent but rather constrained in some way, then the configuration space would be a manifold of smaller dimension.

Usually, the set of solutions of a system of partial differential equations has the structure of some high dimensional manifold. Understanding the "geometry" of this manifold often gives new insight into the nature of these solutions, and to the actual phenomenon that is modeled by the differential equations, whether it comes from physics, economics, engineering, or any other quantitative science.

A typical problem in geometry is to "classify" all manifolds of a certain type. That is, we first decide which kinds of manifolds we are interested in, then decide when two such manifolds should basically be considered to be the same, or "equivalent", and finally try to determine how many inequivalent types of such manifolds exist. For example, we might be interested in studying surfaces (2-dimensional manifolds) that lie inside the usual 3-dimensional space that we can see, and we might decide that two such surfaces are equivalent if one can be "transformed" into the other by translations or rotations. This is the study of the Riemannian geometry of surfaces immersed in 3-space, and was classically the first subfield of "differential geometry", pioneered by mathematical giants such as Gauss and Riemann in the 1800's.

Today, there are many different subfields of geometry that are actively studied. Here we describe only a few of them:

- Riemannian geometry: This is the study of manifolds equipped with the additional structure of a Riemannian metric, which is a rule for measuring lengths of curves and angles between tangent vectors. A Riemannian manifold has curvature, and it is precisely this curvature that makes the laws of classical Euclidean geometry, that we learn in elementary school, to be different. For example, the sum of the interior angles of a "triangle" on a curved Riemannian manifold can be more or less than π if the curvature is positive or negative, respectively.

- Algebraic geometry: This is the study of algebraic varieties, which are solution sets of systems of polynomial equations. They are sometimes manifolds but also often have "singular points" at which they are not "smooth". Because they are defined algebraically, there are many more tools available from abstract algebra to study them, and conversely many questions in pure algebra can be understood better by reformulating the problem in terms of algebraic geometry. Moreover, one can study varieties over any field, not just the real or complex numbers.

- Symplectic geometry: This is the study of manifolds equipped with an additional structure called a symplectic form. A symplectic form is in some sense (that can be made precise) the opposite of a Riemannian metric, and symplectic manifolds exhibit very different behaviour from Riemannian manifolds. For example, a famous theorem of Darboux says that all symplectic manifolds are "locally" the same, although globally they can be extremely different. Such a theorem is far from true in Riemannian geometry. Symplectic manifolds arise naturally in physical systems from classical mechanics, and are called "phases spaces" in physics. This branch of geometry is very topological in nature.

- Complex geometry: This is the study of manifolds which locally "look like" ordinary n-dimensional spaces that are modeled on the complex numbers rather than the real numbers. Because the analysis of holomorphic (or complex-analytic) functions is much more rigid than the real case (for example not all real smooth functions are real-analytic) there are many fewer "types" of complex manifolds, and there has been more success in (at least partial) classifications. This field is also very closely related to algebraic geometry.

For example, the field of Kaehler geometry is in some sense the study of manifolds which lie in the intersection of the above four subfields.

Finally, another very important area of geometry is the study of connections (and their curvature) on vector bundles, also commonly called "gauge theory". This field was independently developed by both physicists and mathematicians around the 1950's. When the two camps finally got together in the 1970's to communicate, led by renowned figures such as Atiyah, Bott, Singer, and Witten, there resulted a spectacular succession of important new advances in both fields. Some of these accomplishments include the existence of "exotic" 4-dimensional manifolds and the discovery of new invariants that distinguish different types of spaces.

Euclidean Geometry

Euclidean geometry is the study of plane and solid figures on the basis of axioms and theorems employed by the Greek mathematician Euclid. In its rough outline, Euclidean geometry is the plane and solid geometry commonly taught in secondary schools. Indeed, until the second half of the 19th century, when non-Euclidean geometries attracted the attention of mathematicians, geometry meant Euclidean geometry. It is the most typical expression of general mathematical thinking. Rather than the memorization of simple algorithms to solve equations by rote, it demands true insight into the subject, clever ideas for applying theorems in special situations, an ability to generalize from known facts, and an insistence on the importance of proof. In Euclid's great work, the Elements, the only tools employed for geometrical constructions were the ruler and the compass—a restriction retained in elementary Euclidean geometry to this day.

In its rigorous deductive organization, the Elements remained the very model of scientific exposition until the end of the 19th century, when the German mathematician David Hilbert wrote his famous Foundations of Geometry. The modern version of Euclidean geometry is the theory of Euclidean (coordinate) spaces of multiple dimensions, where distance is measured by a suitable generalization of the Pythagorean theorem.

Fundamentals

Euclid realized that a rigorous development of geometry must start with the foundations. Hence, he began the Elements with some undefined terms, such as "a point is that which has no part" and "a line is a length without breadth." Proceeding from these terms, he defined further ideas such as angles, circles, triangles, and various other polygons and figures. For example, an angle was defined as the inclination of two straight lines, and a circle was a plane figure consisting of all points that have a fixed distance (radius) from a given centre.

As a basis for further logical deductions, Euclid proposed five common notions, such as "things equal to the same thing are equal," and five unprovable but intuitive principles known variously as postulates or axioms. Stated in modern terms, the axioms are as follows:

- Given two points, there is a straight line that joins them.

- A straight line segment can be prolonged indefinitely.

- A circle can be constructed when a point for its centre and a distance for its radius are given.

- All right angles are equal.

- If a straight line falling on two straight lines makes the interior angles on the same side less than two right angles, the two straight lines, if produced indefinitely, will meet on that side on which the angles are less than the two right angles.

Hilbert refined axioms are as follows:

- For any two different points, (a) there exists a line containing these two points, and (b) this line is unique.

- For any line L and point p not on L, (a) there exists a line through p not meeting L, and (b) this line is unique.

The fifth axiom became known as the "parallel postulate," since it provided a basis for the uniqueness of parallel lines. It also attracted great interest because it seemed less intuitive or self-evident than the others. In the 19th century, Carl Friedrich Gauss, János Bolyai, and Nikolay Lobachevsky all began to experiment with this postulate, eventually arriving at new, non-Euclidean, geometries. All five axioms provided the basis for numerous provable statements, or theorems, on which Euclid built his geometry.

Plane Geometry

Congruence of Triangles

Two triangles are said to be congruent if one can be exactly superimposed on the other by a rigid motion, and the congruence theorems specify the conditions under which this can occur. The first such theorem is the side-angle-side (SAS) theorem: If two sides and the included angle of one triangle are equal to two sides and the included angle of another triangle, the triangles are congruent. Following this, there are corresponding angle-side-angle (ASA) and side-side-side (SSS) theorems.

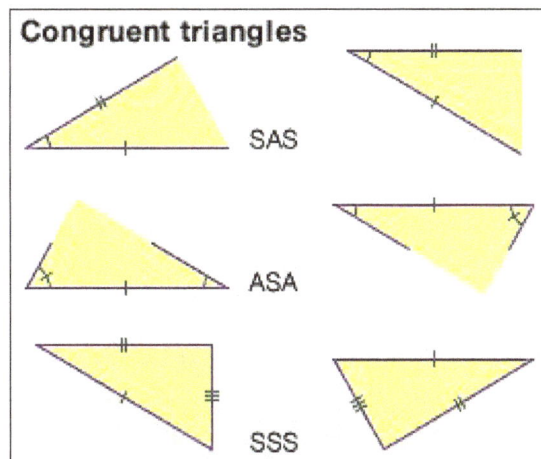

The figure illustrates the three basic theorems that triangles are congruent (of equal shape and size) if: two sides and the included angle are equal (SAS); two angles and the included side are equal (ASA); or all three sides are equal (SSS).

The first very useful theorem derived from the axioms is the basic symmetry property of isosceles triangles—i.e., that two sides of a triangle are equal if and only if the angles opposite them are equal. Euclid's proof of this theorem was once called Pons Asinorum ("Bridge of Asses"), supposedly because mediocre students could not proceed across it to the farther reaches of geometry. The Bridge of Asses opens the way to various theorems on the congruence of triangles.

The parallel postulate is fundamental for the proof of the theorem that the sum of the angles of a triangle is always 180 degrees. A simple proof of this theorem was attributed to the Pythagoreans.

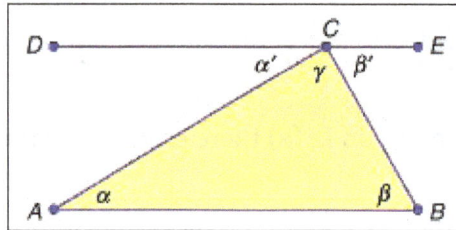

Proof that the sum of the angles in a triangle is 180 degrees. According to an ancient theorem, a transversal through two parallel lines (DE and AB in the figure) forms several equal angles, such as the alternating angles α/α' and β/β', labeled in the figure. By definition, the three angles α', γ, and β' on the line DE must sum to 180 degrees. Since α = α' and β = β', the sum of the angles in the triangle (α, β, and γ) is also 180 degrees.

Similarity of Triangles

As indicated above, congruent figures have the same shape and size. Similar figures, on the other hand, have the same shape but may differ in size. Shape is intimately related to the notion of proportion, as ancient Egyptian artisans observed long ago. Segments of lengths a, b, c, and d are said to be proportional if a:b = c:d (read, a is to b as c is to d; in older notation a:b::c:d). The fundamental theorem of similarity states that a line segment splits two sides of a triangle into proportional segments if and only if the segment is parallel to the triangle's third side.

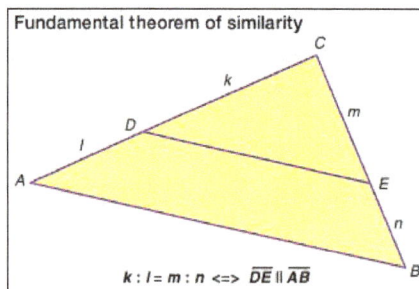

Fundamental theorem of similarity

$k : l = m : n \Leftrightarrow \overline{DE} \parallel \overline{AB}$

The formula in the figure reads k is to l as m is to n if and only if line DE is parallel to line AB. This theorem then enables one to show that the small and large triangles are similar.

The similarity theorem may be reformulated as the AAA (angle-angle-angle) similarity theorem: two triangles have their corresponding angles equal if and only if their corresponding sides are proportional. Two similar triangles are related by a scaling (or similarity) factors: if the first triangle has sides a, b, and c, then the second one will have sides sa, sb, and sc. In addition to the ubiquitous use of scaling factors on construction plans and geographic maps, similarity is fundamental to trigonometry.

Areas

Just as a segment can be measured by comparing it with a unit segment, the area of a polygon or other plane figure can be measured by comparing it with a unit square. The common formulas for calculating areas reduce this kind of measurement to the measurement of certain suitable lengths. The simplest case is a rectangle with sides a and b, which has area ab. By putting a triangle into an appropriate rectangle, one can show that the area of the triangle is half the product of the length of one of its bases and its corresponding height— $bh/2$. One can then compute the area of a general polygon by dissecting it into triangular regions. If a triangle has area A, a similar triangle with a scaling factor of s will have an area of $s^2 A$.

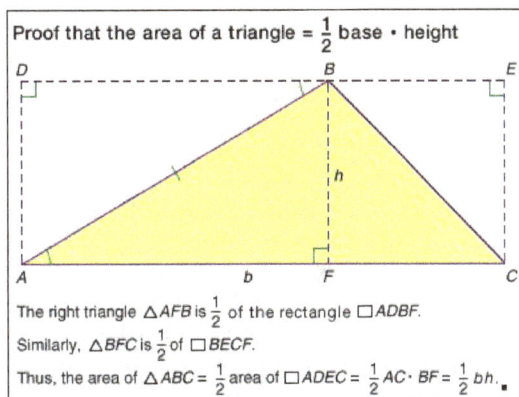

Proof that the area of a triangle = $\frac{1}{2}$ base · height

The right triangle $\triangle AFB$ is $\frac{1}{2}$ of the rectangle $\square ADBF$.

Similarly, $\triangle BFC$ is $\frac{1}{2}$ of $\square BECF$.

Thus, the area of $\triangle ABC = \frac{1}{2}$ area of $\square ADEC = \frac{1}{2} AC \cdot BF = \frac{1}{2} bh$.

Pythagorean Theorem

For a triangle $\triangle ABC$ the Pythagorean theorem has two parts: (1) if $\angle ACB$ is a right angle, then $a^2 + b^2 = c^2$; (2) if $a^2 + b^2 = c^2$, then $\angle ACB$ is a right angle. For an arbitrary triangle, the Pythagorean theorem is generalized to the law of cosines: $a^2 + b^2 = c^2 - 2ab\cos(\angle ACB)$. When $\angle ACB$ is 90 degrees, this reduces to the Pythagorean theorem because $\cos(90°) = 0$.

Since Euclid, a host of professional and amateur mathematicians (even U.S. President James Garfield) have found more than 300 distinct proofs of the Pythagorean theorem. Despite its antiquity, it remains one of the most important theorems in mathematics. It enables one to calculate distances or, more important, to define distances in situations far more general than elementary geometry. For example, it has been generalized to multidimensional vector spaces.

Circles

A chord AB is a segment in the interior of a circle connecting two points (A and B) on the circumference. When a chord passes through the circle's centre, it is a diameter, d. The circumference of a circle is given by πd, or $2\pi r$ where r is the radius of the circle; the area of a circle is πr^2. In each case, π is the same constant (3.14159...). The Greek mathematician Archimedes used the method of exhaustion to obtain upper and lower bounds for π by circumscribing and inscribing regular polygons about a circle.

A semicircle has its end points on a diameter of a circle. Thales is generally credited with having proved that any angle inscribed in a semicircle is a right angle; that is, for any point C on the semicircle with diameter AB, $\angle ACB$ will always be 90 degrees. Another important theorem states that for any chord AB in a circle, the angle subtended by any point on the same semiarc of the circle will be invariant. Slightly modified, this means that in a circle, equal chords determine equal angles, and vice versa.

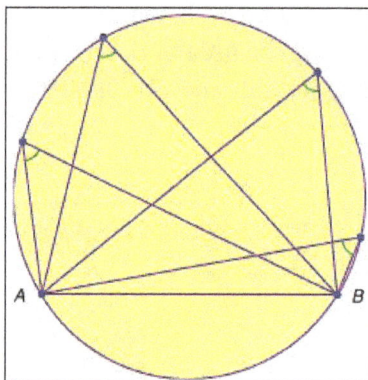

Thales of Miletus is generally credited with giving the first proof that for any chord AB in a circle, all of the angles subtended by points anywhere on the same semiarc of the circle will be equal.

Summarizing the above material, the five most important theorems of plane Euclidean geometry are: the sum of the angles in a triangle is 180 degrees, the Bridge of Asses, the fundamental theorem of similarity, the Pythagorean theorem, and the invariance of angles subtended by a chord in a circle. Most of the more advanced theorems of plane Euclidean geometry are proved with the help of these theorems.

Regular Polygons

A polygon is called regular if it has equal sides and angles. Thus, a regular triangle is an equilateral triangle, and a regular quadrilateral is a square. A general problem since antiquity has been the problem of constructing a regular n-gon, for different n, with only ruler and compass. For example, Euclid constructed a regular pentagon by applying the above-mentioned five important theorems in an ingenious combination.

Techniques, such as bisecting the angles of known constructions, exist for constructing regular n-gons for many values, but none is known for the general case. In 1797, following centuries without any progress, Gauss surprised the mathematical community by discovering a construction for the 17-gon. More generally, Gauss was able to show that for a prime number p, the regular p-gon is constructible if and only if p is a "Fermat prime": $p = F(k) = 2^{2^k} + 1$. Because it is not known in general which $F(k)$ are prime, the construction problem for regular n-gons is still open.

Three other unsolved construction problems from antiquity were finally settled in the 19th century by applying tools not available to the Greeks. Comparatively simple algebraic methods showed that it is not possible to trisect an angle with ruler and compass or to construct a cube with a volume double that of a given cube. Showing that it is not possible to square a circle (i.e., to construct a square equal in area to a given circle by the same means), however, demanded deeper insights into the nature of the number π.

Conic Sections and Geometric Art

The most advanced part of plane Euclidean geometry is the theory of the conic sections (the ellipse, the parabola, and the hyperbola). Much as the Elements displaced all other introductions to geometry, the Conics of Apollonius of Perga, known by his contemporaries as "the Great Geometer," was for many centuries the definitive treatise on the subject.

Medieval Islamic artists explored ways of using geometric figures for decoration. For example, the decorations of the Alhambra of Granada, Spain, demonstrate an understanding of all 17 of the different "Wallpaper groups" that can be used to tile the plane. In the 20th century, internationally renowned artists such as Josef Albers, Max Bill, and Sol LeWitt were inspired by motifs from Euclidean geometry.

Solid Geometry

The most important difference between plane and solid Euclidean geometry is that human beings can look at the plane "from above," whereas three-dimensional space cannot be looked at "from outside." Consequently, intuitive insights are more difficult to obtain for solid geometry than for plane geometry.

Some concepts, such as proportions and angles, remain unchanged from plane to solid geometry. For other familiar concepts, there exist analogies—most noticeably, volume for area and three-dimensional shapes for two-dimensional shapes (sphere for circle, tetrahedron for triangle, box for rectangle). However, the theory of tetrahedra is not nearly as rich as it is for triangles. Active research in higher-dimensional Euclidean geometry includes convexity and sphere packings and their applications in cryptology and crystallography.

Volume

In plane geometry the area of any polygon can be calculated by dissecting it into triangles. A similar procedure is not possible for solids. In 1901 the German mathematician Max Dehn showed that there exist a cube and a tetrahedron of equal volume that cannot be dissected and rearranged into each other. This means that calculus must be used to calculate volumes for even many simple solids such as pyramids.

Regular Solids

Regular polyhedra are the solid analogies to regular polygons in the plane. Regular polygons are defined as having equal (congruent) sides and angles. In analogy, a solid is called regular if its faces are congruent regular polygons and its polyhedral angles (angles at which the faces meet) are congruent. This concept has been generalized to higher-dimensional (coordinate) Euclidean spaces.

Whereas in the plane there exist (in theory) infinitely many regular polygons, in three-dimensional space there exist exactly five regular polyhedra. These are known as the Platonic solids: the tetrahedron, or pyramid, with 4 triangular faces; the cube, with 6 square faces; the octahedron, with 8 equilateral triangular faces; the dodecahedron, with 12 pentagonal faces; and the icosahedron, with 20 equilateral triangular faces.

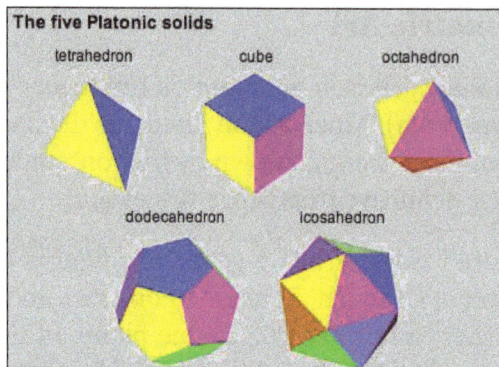

The five Platonic solids: These are the only geometric solids
whose faces are composed of regular, identical polygons.

In four-dimensional space there exist exactly six regular polytopes, five of them generalizations from three-dimensional space. In any space of more than four dimensions, there exist exactly three regular polytopes—the generalizations of the tetrahedron, the cube, and the octahedron.

Calculating Areas and Volumes

Mathematical formulas			
	Shape	Action	Formula
Circumference	Circle	Multiply diameter by π	πd
Area	Circle	Multiply radius squared by π	$πr^2$
	Rectangle	Multiply height by length	hl
	Sphere Surface	Multiply radius squared by π by 4	$4πr^2$
	Square	Length of one side squared	s^2
	Trapezoid	Parallel side length A + parallel side length B multiplied by height and divided by 2	(A + B)h/2
	Triangle	Multiply base by height and divide by 2	hb/2
Volume	Cone	Multiply base radius squared by π by height and divide by 3	$br^2πh/3$
	Cube	Length of one edge cubed	a^3
	Cylinder	Multiply base radius squared by π by height	$br^2πh$
	Pyramid	Multiply base length by base width by height and divide by 3	lwh/3
	Sphere	Multiply radius cubed by π by 4 and divide by 3	$4πr^3/3$

Analytic Geometry

In classical mathematics, analytic geometry, also known as coordinate geometry or Cartesian geometry, is the study of geometry using a coordinate system. This contrasts with synthetic geometry.

Analytic geometry is widely used in physics and engineering, and also in aviation, rocketry, space science, and spaceflight. It is the foundation of most modern fields of geometry, including algebraic, differential, discrete and computational geometry.

Usually the Cartesian coordinate system is applied to manipulate equations for planes, straight lines, and squares, often in two and sometimes in three dimensions. Geometrically, one studies the Euclidean plane (two dimensions) and Euclidean space (three dimensions). As taught in school books, analytic geometry can be explained more simply: it is concerned with defining and representing geometrical shapes in a numerical way and extracting numerical information from shapes' numerical definitions and representations. That the algebra of the real numbers can be employed to yield results about the linear continuum of geometry relies on the Cantor–Dedekind axiom.

Coordinates

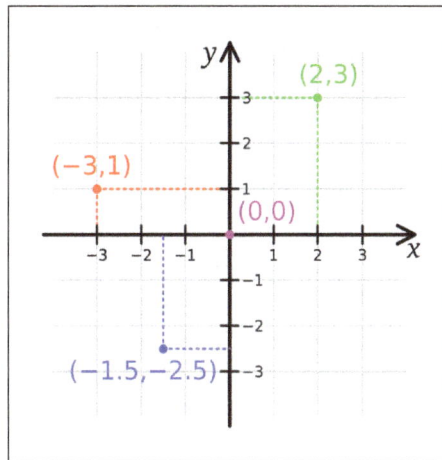

Illustration of a Cartesian coordinate plane. Four points are marked and labeled with their coordinates: (2,3) in green, (−3,1) in red, (−1.5,−2.5) in blue, and the origin (0,0) in purple.

In analytic geometry, the plane is given a coordinate system, by which every point has a pair of real number coordinates. Similarly, Euclidean space is given coordinates where every point has three coordinates. The value of the coordinates depends on the choice of the initial point of origin. There are a variety of coordinate systems used, but the most common are the following:

Cartesian Coordinates (in a Plane or Space)

The most common coordinate system to use is the Cartesian coordinate system, where each point has an x-coordinate representing its horizontal position, and a y-coordinate representing its vertical position. These are typically written as an ordered pair (x,y). This system can also be used for three-dimensional geometry, where every point in Euclidean space is represented by an ordered triple of coordinates (x,y,z).

Polar Coordinates (in a Plane)

In polar coordinates, every point of the plane is represented by its distance r from the origin and its angle θ from the polar axis.

Cylindrical Coordinates (in a Space)

In cylindrical coordinates, every point of space is represented by its height z, its radius r from the z-axis and the angle θ its projection on the xy-plane makes with respect to the horizontal axis.

Spherical Coordinates (in a Space)

In spherical coordinates, every point in space is represented by its distance ρ from the origin, the angle θ its projection on the xy-plane makes with respect to the horizontal axis, and the angle φ that it makes with respect to the z-axis. The names of the angles are often reversed in physics.

Equations and Curves

In analytic geometry, any equation involving the coordinates specifies a subset of the plane, namely the solution set for the equation, or locus. For example, the equation $y = x$ corresponds to the set of all the points on the plane whose x-coordinate and y-coordinate are equal. These points form a line, and $y = x$ is said to be the equation for this line. In general, linear equations involving x and y specify lines, quadratic equations specify conic sections, and more complicated equations describe more complicated figures.

Usually, a single equation corresponds to a curve on the plane. This is not always the case: the trivial equation $x = x$ specifies the entire plane, and the equation $x^2 + y^2 = 0$ specifies only the single point (0, 0). In three dimensions, a single equation usually gives a surface, and a curve must be specified as the intersection of two surfaces, or as a system of parametric equations. The equation $x^2 + y^2 = r^2$ is the equation for any circle centered at the origin (0, 0) with a radius of r.

Lines and Planes

Lines in a Cartesian plane or, more generally, in affine coordinates, can be described algebraically by *linear* equations. In two dimensions, the equation for non-vertical lines is often given in the *slope-intercept form*:

$$y = mx + b$$

where:

   m is the slope or gradient of the line.

   b is the y-intercept of the line.

   x is the independent variable of the function $y = f(x)$.

In a manner analogous to the way lines in a two-dimensional space are described using a point-slope form for their equations, planes in a three dimensional space have a natural description using a point in the plane and a vector orthogonal to it (the normal vector) to indicate its "inclination".

Specifically, let $\mathbf{r}_0$ be the position vector of some point $P_0 = (x_0, y_0, z_0)$, and let $\mathbf{n} = (a, b, c)$ be a nonzero vector. The plane determined by this point and vector consists of those points P, with position vector $\mathbf{r}$, such that the vector drawn from P_0 to P is perpendicular to $\mathbf{n}$. Recalling that two vectors are perpendicular if and only if their dot product is zero, it follows that the desired plane can be described as the set of all points $\mathbf{r}$ such that:

$$\mathbf{n} \cdot (\mathbf{r} - \mathbf{r}_0) = 0.$$

(The dot here means a dot product, not scalar multiplication.) Expanded, this becomes:

$$a(x - x_0) + b(y - y_0) + c(z - z_0) = 0,$$

which is the *point-normal* form of the equation of a plane. This is just a linear equation:

$$ax + by + cz + d = 0, \text{ where } d = -(ax_0 + by_0 + cz_0).$$

Conversely, it is easily shown that if a,b,c and d are constants and $a,b,$ and c are not all zero, then the graph of the equation:

$$ax + by + cz + d = 0,$$

is a plane having the vector $\mathbf{n} = (a,b,c)$ as a normal. This familiar equation for a plane is called the *general form* of the equation of the plane.

In three dimensions, lines can *not* be described by a single linear equation, so they are frequently described by parametric equations:

$$x = x_0 + at$$
$$y = y_0 + bt$$
$$z = z_0 + ct$$

where:

x, y, and z are all functions of the independent variable t which ranges over the real numbers.

(x_0, y_0, z_0) is any point on the line.

$a, b,$ and c are related to the slope of the line, such that the vector (a, b, c) is parallel to the line.

Conic Sections

In the Cartesian coordinate system, the graph of a quadratic equation in two variables is always a conic section – though it may be degenerate, and all conic sections arise in this way. The equation will be of the form:

$$Ax^2 + Bxy + Cy^2 + Dx + Ey + F = 0 \text{ with } A, B, C \text{ not all zero.}$$

As scaling all six constants yields the same locus of zeros, one can consider conics as points in the five-dimensional projective space.

The conic sections described by this equation can be classified using the discriminant:

$$B^2 - 4AC.$$

If the conic is non-degenerate, then:

- If $B^2 - 4AC < 0$, the equation represents an ellipse:

  ◦ If $A = C$ and $B = 0$, the equation represents a circle, which is a special case of an ellipse.

- If $B^2 - 4AC = 0$, the equation represents a parabola.

- If $B^2 - 4AC > 0$, the equation represents a hyperbola:

  ◦ If we also have $A + C = 0$, the equation represents a rectangular hyperbola.

Quadric Surfaces

A quadric, or quadric surface, is a 2-dimensional surface in 3-dimensional space defined as the locus of zeros of a quadratic polynomial. In coordinates x_1, x_2, x_3, the general quadric is defined by the algebraic equation,

$$\sum_{i,j=1}^{3} x_i Q_{ij} x_j + \sum_{i=1}^{3} P_i x_i + R = 0.$$

Quadric surfaces include ellipsoids (including the sphere), paraboloids, hyperboloids, cylinders, cones, and planes.

Distance and Angle

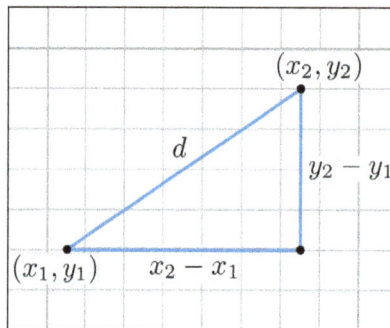

The distance formula on the plane follows from the Pythagorean theorem.

In analytic geometry, geometric notions such as distance and angle measure are defined using formulas. These definitions are designed to be consistent with the underlying Euclidean geometry. For example, using Cartesian coordinates on the plane, the distance between two points (x_1, y_1) and (x_2, y_2) is defined by the formula:

$$d = \sqrt{(x_2 - x_1)^2 + (y_2 - y_1)^2},$$

which can be viewed as a version of the Pythagorean theorem. Similarly, the angle that a line makes with the horizontal can be defined by the formula:

$$\theta = \arctan(m),$$

where m is the slope of the line.

In three dimensions, distance is given by the generalization of the Pythagorean theorem:

$$d = \sqrt{(x_2 - x_1)^2 + (y_2 - y_1)^2 + (z_2 - z_1)^2},$$

while the angle between two vectors is given by the dot product. The dot product of two Euclidean vectors A and B is defined by:

$$\mathbf{A} \cdot \mathbf{B} \stackrel{\text{def}}{=} \|\mathbf{A}\| \|\mathbf{B}\| \cos\theta,$$

where θ is the angle between A and B.

Transformations

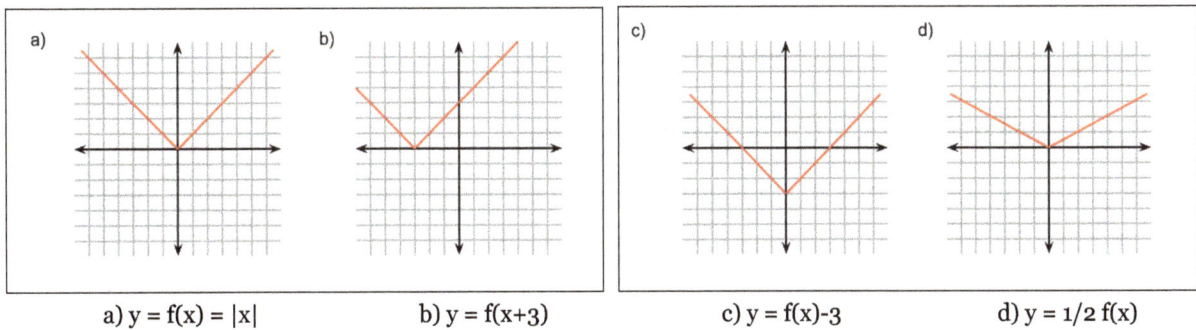

| a) y = f(x) = |x| | b) y = f(x+3) | c) y = f(x)-3 | d) y = 1/2 f(x) |

Transformations are applied to a parent function to turn it into a new function with similar characteristics.

The graph of $R(x, y)$ is changed by standard transformations as follows:

- Changing x to $x - h$ moves the graph to the right h units.

- Changing y to $y - k$ moves the graph up k units.

- Changing x to x/b stretches the graph horizontally by a factor of b. (think of the x as being dilated).

- Changing y to y/a stretches the graph vertically.

- Changing x to $x\cos A + y\sin A$ and changing y to $-x\sin A + y\cos A$ rotates the graph by an angle A.

There are other standard transformation not typically studied in elementary analytic geometry because the transformations change the shape of objects in ways not usually considered. Skewing is an example of a transformation not usually considered.

For example, the parent function $y = 1/x$ has a horizontal and a vertical asymptote, and occupies the first and third quadrant, and all of its transformed forms have one horizontal and vertical asymptote, and occupies either the 1st and 3rd or 2nd and 4th quadrant. In general, if $y = f(x)$, then it can be transformed into $y = af(b(x - k)) + h$. In the new transformed function, a is the factor that vertically stretches the function if it is greater than 1 or vertically compresses the function if it is less than 1, and for negative a values, the function is reflected in the x-axis. The b value compresses the graph of the function horizontally if greater than 1 and stretches the function horizontally if less than 1, and like a, reflects the function in the $y-$axis when it is negative. The k and h values introduce translations, h, vertical, and k horizontal. Positive h and k values mean the function is translated to the positive end of its axis and negative meaning translation towards the negative end.

Transformations can be applied to any geometric equation whether or not the equation represents a function. Transformations can be considered as individual transactions or in combinations.

Suppose that $R(x, y)$ is a relation in the xy plane. For example,

$$x^2 + y^2 - 1 = 0$$

is the relation that describes the unit circle.

Finding Intersections of Geometric Objects

For two geometric objects P and Q represented by the relations $P(x, y)$ and $Q(x, y)$ the intersection is the collection of all points (x, y) which are in both relations.

For example, P might be the circle with radius 1 and center $(0,0)$: $P = \{(x, y) \mid x^2 + y^2 = 1\}$ and Q might be the circle with radius 1 and center $(1,0)$: $Q = \{(x, y) \mid (x-1)^2 + y^2 = 1\}$. The intersection of these two circles is the collection of points which make both equations true. Does the point $(0,0)$ make both equations true? Using $(0,0)$ for (x, y), the equation for Q becomes $(0-1)^2 + 0^2 = 1$ or $(-1)^2 = 1$ which is true, so $(0,0)$ is in the relation Q. On the other hand, still using $(0,0)$ for (x, y) the equation for P becomes $0^2 + 0^2 = 1$ or $0 = 1$ which is false. $(0,0)$ is not in P so it is not in the intersection.

The intersection of P and Q can be found by solving the simultaneous equations:

$$x^2 + y^2 = 1$$
$$(x-1)^2 + y^2 = 1.$$

Traditional methods for finding intersections include substitution and elimination.

Substitution: Solve the first equation for y in terms of x and then substitute the expression for into the second equation:

$$x^2 + y^2 = 1$$
$$y^2 = 1 - x^2.$$

We then substitute this value for y^2 into the other equation and proceed to solve for x:

$$(x-1)^2 + (1 - x^2) = 1$$
$$x^2 - 2x + 1 + 1 - x^2 = 1$$
$$-2x = -1$$
$$x = 1/2.$$

Next, we place this value of x in either of the original equations and solve for y:

$$(1/2)^2 + y^2 = 1$$
$$y^2 = 3/4$$
$$y = \frac{\pm\sqrt{3}}{2}.$$

So our intersection has two points:

$$\left(1/2, \frac{+\sqrt{3}}{2}\right) \text{ and } \left(1/2, \frac{-\sqrt{3}}{2}\right).$$

Elimination: Add (or subtract) a multiple of one equation to the other equation so that one of the variables is eliminated. For our current example, if we subtract the first equation from the second we get $(x-1)^2 - x^2 = 0$. The y^2 in the first equation is subtracted from the y^2 in the second equation leaving no y term. The variable y has been eliminated. We then solve the remaining equation for x, in the same way as in the substitution method:

$$x^2 - 2x + 1 + 1 - x^2 = 1$$
$$-2x = -1$$
$$x = 1/2.$$

We then place this value of x in either of the original equations and solve for y:

$$(1/2)^2 + y^2 = 1$$
$$y^2 = 3/4$$
$$y = \frac{\pm\sqrt{3}}{2}.$$

So our intersection has two points:

$$\left(1/2, \frac{+\sqrt{3}}{2}\right) \text{ and } \left(1/2, \frac{-\sqrt{3}}{2}\right).$$

For conic sections, as many as 4 points might be in the intersection.

Finding Intercepts

One type of intersection which is widely studied is the intersection of a geometric object with the x and y coordinate axes.

The intersection of a geometric object and the y-axis is called the y-intercept of the object. The intersection of a geometric object and the x-axis is called the x-intercept of the object.

For the line $y = mx + b$, the parameter b specifies the point where the line crosses the y axis. Depending on the context, either b or the point $(0, b)$ is called the y-intercept.

Tangents and Normals

Tangent Lines and Planes

In geometry, the tangent line (or simply tangent) to a plane curve at a given point is the straight line that "just touches" the curve at that point. Informally, it is a line through a pair of infinitely close points on the curve. More precisely, a straight line is said to be a tangent of a curve $y = f(x)$ at a point $x = c$ on the curve if the line passes through the point $(c, f(c))$ on the curve and has slope $f'(c)$ where f is the derivative of f. A similar definition applies to space curves and curves in n-dimensional Euclidean space.

As it passes through the point where the tangent line and the curve meet, called the point of tangency, the tangent line is "going in the same direction" as the curve, and is thus the best straight-line approximation to the curve at that point.

Similarly, the tangent plane to a surface at a given point is the plane that "just touches" the surface at that point. The concept of a tangent is one of the most fundamental notions in differential geometry and has been extensively generalized.

Normal Line and Vector

In geometry, a normal is an object such as a line or vector that is perpendicular to a given object. For example, in the two-dimensional case, the normal line to a curve at a given point is the line perpendicular to the tangent line to the curve at the point.

In the three-dimensional case a surface normal, or simply normal, to a surface at a point P is a vector that is perpendicular to the tangent plane to that surface at P. The word "normal" is also used as an adjective: a line normal to a plane, the normal component of a force, the normal vector, etc. The concept of normality generalizes to orthogonality.

Non-euclidean Geometry

Non-Euclidean geometry, literally any geometry that is not the same as Euclidean geometry. Although the term is frequently used to refer only to hyperbolic geometry, common usage includes those few geometries (hyperbolic and spherical) that differ from but are very close to Euclidean geometry.

Comparison of Euclidean, spherical, and hyperbolic geometries	
Given a line and a point not on the line, there exist(s) _____ through the given point and parallel to the given line.	
a) exactly one line	(Euclidean)
b) no lines	(spherical)
c) infinitely many lines	(hyperbolic)

Euclid's fifth postulate is _____.		
	a) true	(Euclidean)
	b) false	(spherical)
	c) false	(hyperbolic)
The sum of the interior angles of a triangle _____ 180 degrees.		
	a) =	(Euclidean)
	b) >	(spherical)
	c) <	(hyperbolic)

The non-Euclidean geometries developed along two different historical threads. The first thread started with the search to understand the movement of stars and planets in the apparently hemispherical sky. For example, Euclid wrote about spherical geometry in his astronomical work Phaenomena. In addition to looking to the heavens, the ancients attempted to understand the shape of the Earth and to use this understanding to solve problems in navigation over long distances (and later for large-scale surveying). These activities are aspects of spherical geometry.

The second thread started with the fifth ("parallel") postulate in Euclid's Elements:

- If a straight line falling on two straight lines makes the interior angles on the same side less than two right angles, the two straight lines, if produced indefinitely, will meet on that side on which the angles are less than the two right angles.

For 2,000 years following Euclid, mathematicians attempted either to prove the postulate as a theorem (based on the other postulates) or to modify it in various ways. These attempts culminated when the Russian Nikolay Lobachevsky and the Hungarian János Bolyai independently published a description of a geometry that, except for the parallel postulate, satisfied all of Euclid's postulates and common notions. It is this geometry that is called hyperbolic geometry.

Spherical Geometry

From early times, people noticed that the shortest distance between two points on Earth were great circle routes. For example, the Greek astronomer Ptolemy wrote in Geography:

> It has been demonstrated by mathematics that the surface of the land and water is in its entirety a sphere and that any plane which passes through the centre makes at its surface, that is, at the surface of the Earth and of the sky, great circles.

Great circles are the "straight lines" of spherical geometry. This is a consequence of the properties of a sphere, in which the shortest distances on the surface are great circle routes. Such curves are said to be "intrinsically" straight. Note, however, that intrinsically straight and shortest are not necessarily identical, as shown in the figure. Three intersecting great circle arcs form a spherical triangle; while a spherical triangle must be distorted to fit on another sphere with a different radius, the difference is only one of scale. In differential geometry, spherical geometry is described as the geometry of a surface with constant positive curvature.

There are many ways of projecting a portion of a sphere, such as the surface of the Earth, onto a plane. These are known as maps or charts and they must necessarily distort distances and either area or angles. Cartographers' need for various qualities in map projections gave an early impetus to the study of spherical geometry.

Elliptic geometry is the term used to indicate an axiomatic formalization of spherical geometry in which each pair of antipodal points is treated as a single point. An intrinsic analytic view of spherical geometry was developed in the 19th century by the German mathematician Bernhard Riemann; usually called the Riemann sphere, it is studied in university courses on complex analysis. Some texts call this (and therefore spherical geometry) Riemannian geometry, but this term more correctly applies to a part of differential geometry that gives a way of intrinsically describing any surface.

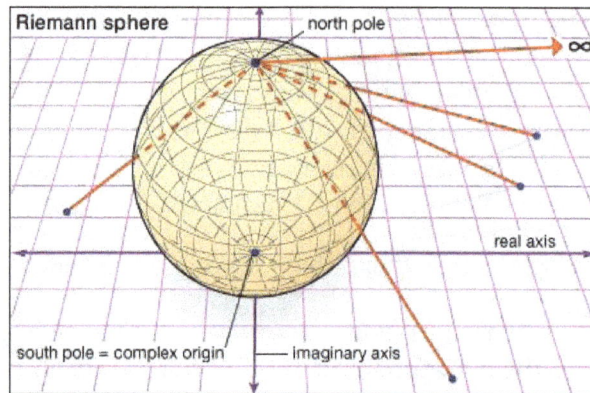

Hyperbolic Geometry

The first description of hyperbolic geometry was given in the context of Euclid's postulates, and it was soon proved that all hyperbolic geometries differ only in scale (in the same sense that spheres only differ in size). In the mid-19th century it was shown that hyperbolic surfaces must have constant negative curvature. However, this still left open the question of whether any surface with hyperbolic geometry actually exists.

Hyperbolic plane, designed and crocheted by Daina Taimina.

In 1868 the Italian mathematician Eugenio Beltrami described a surface, called the pseudosphere, that has constant negative curvature. However, the pseudosphere is not a complete model for hyperbolic geometry, because intrinsically straight lines on the pseudosphere may intersect themselves and cannot be continued past the bounding circle (neither of which is true in hyperbolic geometry). In 1901 the German mathematician David Hilbert proved that it is impossible to define a complete hyperbolic surface using real analytic functions (essentially, functions that can be expressed in terms of ordinary formulas). In those days, a surface always meant one defined by real analytic functions, and so the search was abandoned. However, in 1955 the Dutch mathematician Nicolaas Kuiper proved the existence of a complete hyperbolic surface, and in the 1970s the American mathematician William Thurston described the construction of a hyperbolic surface. Such a surface, as shown in the figure, can also be crocheted.

In the 19th century, mathematicians developed three models of hyperbolic geometry that can now be interpreted as projections (or maps) of the hyperbolic surface. Although these models all suffer from some distortion—similar to the way that flat maps distort the spherical Earth—they are useful individually and in combination as aides to understand hyperbolic geometry. In 1869–71 Beltrami and the German mathematician Felix Klein developed the first complete model of hyperbolic geometry (and first called the geometry "hyperbolic"). In the Klein-Beltrami model, the hyperbolic surface is mapped to the interior of a circle, with geodesics in the hyperbolic surface corresponding to chords in the circle. Thus, the Klein-Beltrami model preserves "straightness" but at the cost of distorting angles. About 1880 the French mathematician Henri Poincaré developed two more models. In the Poincaré disk model, the hyperbolic surface is mapped to the interior of a circular disk, with hyperbolic geodesics mapping to circular arcs (or diameters) in the disk that meet the bounding circle at right angles. In the Poincaré upper half-plane model, the hyperbolic surface is mapped onto the half-plane above the x-axis, with hyperbolic geodesics mapped to semicircles (or vertical rays) that meet the x-axis at right angles. Both Poincaré models distort distances while preserving angles as measured by tangent lines.

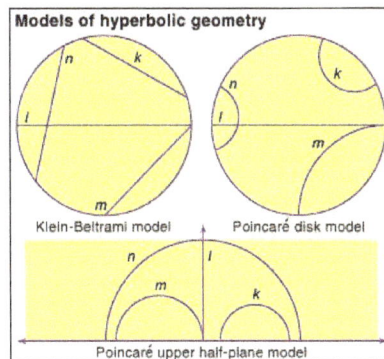

Models of hyperbolic geometry

Klein-Beltrami model

Poincaré disk model

Poincaré upper half-plane model

In the Klein-Beltrami model for the hyperbolic plane, the shortest paths, or geodesics, are chords (several examples, labeled k, l, m, n, are shown). In the Poincaré disk model, geodesics are portions of circles that intersect the boundary of the disk at right angles; and in the Poincaré upper half-plane model, geodesics are semicircles with their centres on the boundary.

Differential Geometry

Differential geometry is a branch of mathematics that studies the geometry of curves, surfaces, and manifolds (the higher-dimensional analogs of surfaces). The discipline owes its name to its use of ideas and techniques from differential calculus, though the modern subject often uses algebraic and purely geometric techniques instead. Although basic definitions, notations, and analytic descriptions vary widely, the following geometric questions prevail: How does one measure the curvature of a curve within a surface (intrinsic) versus within the encompassing space (extrinsic)? How can the curvature of a surface be measured? What is the shortest path within a surface between two points on the surface? How is the shortest path on a surface related to the concept of a straight line?

While curves had been studied since antiquity, the discovery of calculus in the 17th century opened up the study of more complicated plane curves—such as those produced by the French mathematician René Descartes with his "compass". In particular, integral calculus led to general solutions of the ancient problems of finding the arc length of plane curves and the area of plane figures. This in turn opened the stage to the investigation of curves and surfaces in space—an investigation that was the start of differential geometry.

Some of the fundamental ideas of differential geometry can be illustrated by the strake, a spiraling strip often designed by engineers to give structural support to large metal cylinders such as smokestacks. A strake can be formed by cutting an annular strip (the region between two concentric circles) from a flat sheet of steel and then bending it into a helix that spirals around the cylinder, as illustrated in the figure. What should the radius r of the annulus be to produce the best fit? Differential geometry supplies the solution to this problem by defining a precise measurement for the curvature of a curve; then r can be adjusted until the curvature of the inside edge of the annulus matches the curvature of the helix.

An important question remains - can the annular strip be bent, without stretching, so that it forms a strake around the cylinder? In particular, this means that distances measured along the surface (intrinsic) are unchanged. Two surfaces are said to be isometric if one can be bent (or transformed) into the other without changing intrinsic distances. (For example, because a sheet of paper can be rolled into a tube without stretching, the sheet and tube are "locally" isometric—only locally because new, and possibly shorter, routes are created by connecting the two edges of the paper.) Thus, the second question becomes: Are the annular strip and the strake isometric? To answer this and similar questions, differential geometry developed the notion of the curvature of a surface.

Curvature of Curves

Although mathematicians from antiquity had described some curves as curving more than others and straight lines as not curving at all, it was the German mathematician Gottfried Leibniz who, in 1686, first defined the curvature of a curve at each point in terms of the circle that best approximates the curve at that point. Leibniz named his approximating circle (as shown in the figure) the osculating circle, from the Latin osculare ("to kiss"). He then defined the curvature of the curve (and the circle) as 1/r, where r is the radius of the osculating circle. As a curve becomes straighter, a circle with a larger radius must be used to approximate it, and so the resulting curvature decreases. In the limit, a straight line is said to be equivalent to a circle of infinite radius and its curvature defined as zero everywhere.

The only curves in ordinary Euclidean space with constant curvature are straight lines, circles, and helices. In practice, curvature is found with a formula that gives the rate of change, or derivative, of the tangent to the curve as one moves along the curve. This formula was discovered by Isaac Newton and Leibniz for plane curves in the 17th century and by the Swiss mathematician Leonhard Euler for curves in space in the 18th century.

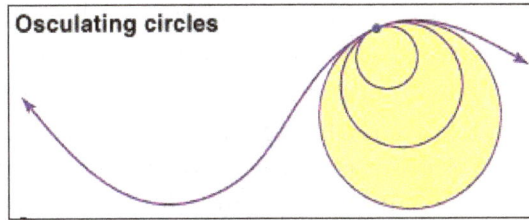

The curvature at each point of a line is defined to be 1/r, where r is the radius of the osculating, or "kissing," circle that best approximates the line at the given point.

With these definitions in place, it is now possible to compute the ideal inner radius r of the annular strip that goes into making the strake shown in the figure. The annular strip's inner curvature $^1/_r$ must equal the curvature of the helix on the cylinder. If R is the radius of the cylinder and H is the height of one turn of the helix, then the curvature of the helix is $4\pi^2 R / [H^2 + (2\pi R)^2]$. For example, if $R = 1$ metre and $H = 10$ metres, then $r = 3.533$ metres.

Curvature of Surfaces

To measure the curvature of a surface at a point, Euler, in 1760, looked at cross sections of the surface made by planes that contain the line perpendicular (or "normal") to the surface at the point. Euler called the curvatures of these cross sections the normal curvatures of the surface at the point. For example, on a right cylinder of radius r, the vertical cross sections are straight lines and thus have zero curvature; the horizontal cross sections are circles, which have curvature $^1/_r$. The normal curvatures at a point on a surface are generally different in different directions. The maximum and minimum normal curvatures at a point on a surface are called the principal (normal) curvatures, and the directions in which these normal curvatures occur are called the principal directions. Euler proved that for most surfaces where the normal curvatures are not constant (for example, the cylinder), these principal directions are perpendicular to each other. Note that on a sphere all the normal curvatures are the same and thus all are principal curvatures. These principal normal curvatures are a measure of how "curvy" the surface is:

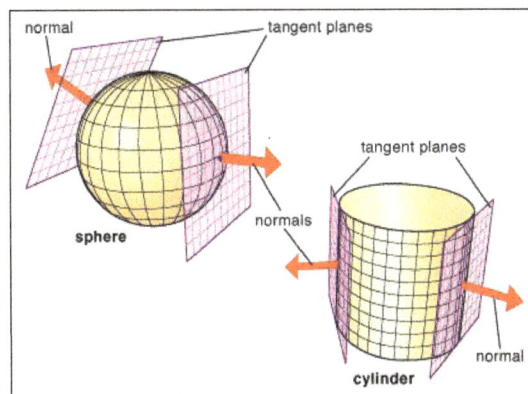

The normal, or perpendicular, at each point of a surface defines the corresponding tangent plane, and vice versa.

The theory of surfaces and principal normal curvatures was extensively developed by French geometers led by Gaspard Monge. It was in an 1827 paper, however, that the German mathematician Carl Friedrich Gauss made the big breakthrough that allowed differential geometry to answer the question raised above of whether the annular strip is isometric to the strake. The Gaussian curvature of a surface at a point is defined as the product of the two principal normal curvatures; it is said to be positive if the principal normal curvatures curve in the same direction and negative if they curve in opposite directions. Normal curvatures for a plane surface are all zero, and thus the Gaussian curvature of a plane is zero. For a cylinder of radius r, the minimum normal curvature is zero (along the vertical straight lines), and the maximum is $1/_r$ (along the horizontal circles). Thus, the Gaussian curvature of a cylinder is also zero.

If the cylinder is cut along one of the vertical straight lines, the resulting surface can be flattened (without stretching) onto a rectangle. In differential geometry, it is said that the plane and cylinder are locally isometric. These are special cases of two important theorems:

- Gauss's Theorem: If two smooth surfaces are isometric, then the two surfaces have the same Gaussian curvature at corresponding points. Athough defined extrinsically, Gaussian curvature is an intrinsic notion.

- Minding's theorem: Two smooth ("cornerless") surfaces with the same constant Gaussian curvature are locally isometric.

As corollaries to these theorems:

- A surface with constant positive Gaussian curvature c has locally the same intrinsic geometry as a sphere of radius $\sqrt{1/_c}$. This is because a sphere of radius r has Gaussian curvature $1/r^2$.

- A surface with constant zero Gaussian curvature has locally the same intrinsic geometry as a plane. Such surfaces are called developable.

- A surface with constant negative Gaussian curvature c has locally the same intrinsic geometry as a hyperbolic plane.

The Gaussian curvature of an annular strip (being in the plane) is constantly zero. So to answer whether or not the annular strip is isometric to the strake, one needs only to check whether a strake has constant zero Gaussian curvature. The Gaussian curvature of a strake is actually negative, hence the annular strip must be stretched—although this can be minimized by narrowing the shapes.

Shortest Paths on a Surface

From an outside, or extrinsic, perspective, no curve on a sphere is straight. Nevertheless, the great circles are intrinsically straight—an ant crawling along a great circle does not turn or curve with respect to the surface. About 1830 the Estonian mathematician Ferdinand Minding defined a curve on a surface to be a geodesic if it is intrinsically straight—that is, if there is no identifiable curvature from within the surface. A major task of differential geometry is to determine the geodesics on a surface. The great circles are the geodesics on a sphere.

A great circle arc that is longer than a half circle is intrinsically straight on the sphere, but it is not the shortest distance between its endpoints. On the other hand, the shortest path in a surface is not always straight, as shown in the figure.

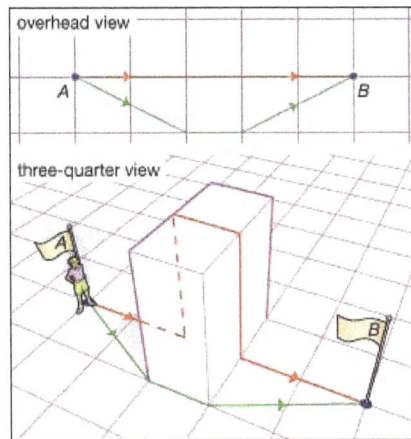

The shaded elevation and the surrounding plane form one continuous surface. Therefore, the red path from A to B that rises over the elevation is intrinsically straight (as viewed from within the surface). However, it is longer than the intrinsically bent green path, demonstrating that an intrinsically straight line is not necessarily the shortest distance between two points.

On a surface which is complete (every geodesic can be extended indefinitely) and smooth, every shortest curve is intrinsically straight and every intrinsically straight curve is the shortest curve between nearby points.

Projective Geometry

Projective geometry is the branch of mathematics that deals with the relationships between geometric figures and the images, or mappings, that result from projecting them onto another surface. Common examples of projections are the shadows cast by opaque objects and motion pictures displayed on a screen.

Projective geometry has its origins in the early Italian Renaissance, particularly in the architectural drawings of Filippo Brunelleschi and Leon Battista Alberti, who invented the method of perspective drawing. By this method, as shown in the figure, the eye of the painter is connected to points on the landscape (the horizontal reality plane, RP) by so-called sight lines. The intersection of these sight lines with the vertical picture plane (PP) generates the drawing. Thus, the reality plane is projected onto the picture plane, hence the name projective geometry.

Although some isolated properties concerning projections were known in antiquity, particularly in the study of optics, it was not until the 17th century that mathematicians returned to the subject. The French mathematicians Girard Desargues and Blaise Pascal took the first significant steps by examining what properties of figures were preserved (or invariant) under perspective mappings. The subject's real importance, however, became clear only after 1800 in the works of several other French mathematicians, notably Jean-Victor Poncelet. In general, by ignoring geometric measurements such as distances and angles, projective geometry enables a clearer understanding of some more generic properties of geometric objects. Such insights have since been incorporated in many more advanced areas of mathematics.

Parallel Lines and the Projection of Infinity

A theorem from Euclid's Elements states that if a line is drawn through a triangle such that it is parallel to one side, then the line will divide the other two sides proportionately; that is, the ratio of segments on each side will be equal. This is known as the proportional segments theorem, or the fundamental theorem of similarity, and for triangle ABC, shown in the diagram, with line segment DE parallel to side AB, the theorem corresponds to the mathematical expression CD/DA = CE/EB.

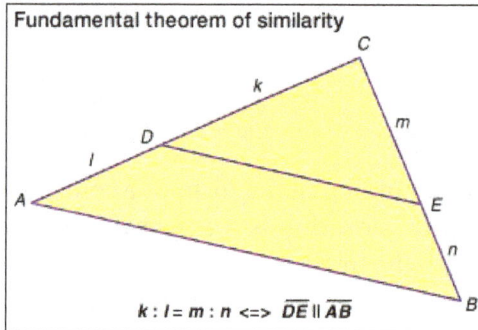

Fundamental theorem of similarity

$k : l = m : n \iff \overline{DE} \parallel \overline{AB}$

The formula in the figure reads k is to l as m is to n if and only if line DE is parallel to line AB. This theorem then enables one to show that the small and large triangles are similar.

Now consider the effect produced by projecting these line segments onto another plane as shown in the figure. The first thing to note is that the projected line segments A′B′ and D′E′ are not parallel; i.e., angles are not preserved. From the point of view of the projection, the parallel lines AB and DE appear to converge at the horizon, or at infinity, whose projection in the picture plane is labeled Ω. It was Desargues who first introduced a single point at infinity to represent the projected intersection of parallel lines. Furthermore, he collected all the points along the horizon in one line at infinity. With the introduction of Ω, the projected figure corresponds to a theorem discovered by Menelaus of Alexandria in the 1st century AD:

$$C'D' / D'A' = C'E' / E'B' \cdot \Omega B' / \Omega A'.$$

Since the factor $\Omega B' / \Omega A'$ corrects for the projective distortion in lengths, Menelaus's theorem can be seen as a projective variant of the proportional segments theorem.

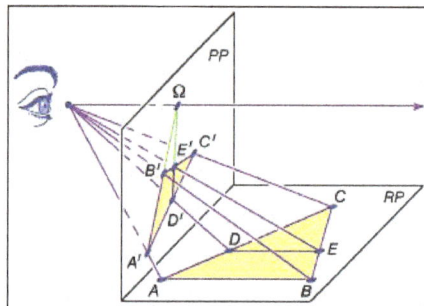

Projective version of the fundamental theorem of similarity. In RP, Euclid's fundamental theorem of similarity states that $CD/DA = CE/EB$. By introducing a scaling factor, the theorem can be saved in RP as $C'D' / D'A' = C'E' / E'B' \cdot \Omega B' / \Omega A'$. Note that while lines AB and DE are parallel in RP, their projections onto PP intersect at the infinitely distant horizon (Ω).

Projective Invariants

With Desargues's provision of infinitely distant points for parallels, the reality plane and the projective plane are essentially interchangeable—that is, ignoring distances and directions (angles), which are not preserved in the projection. Other properties are preserved, however. For instance, two different points have a unique connecting line, and two different lines have a unique point of intersection. Although almost nothing else seems to be invariant under projective mappings, one should note that lines are mapped onto lines. This means that if three points are collinear (share a common line), then the same will be true for their projections. Thus, collinearity is another invariant property. Similarly, if three lines meet in a common point, so will their projections.

The following theorem is of fundamental importance for projective geometry. In its first variant, by Pappus of Alexandria as shown in the figure, it only uses collinearity:

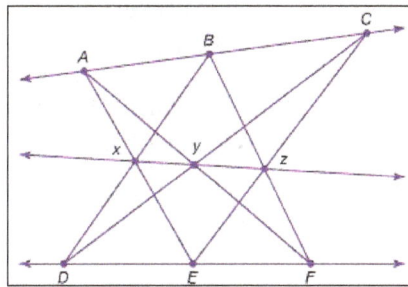

Pappus's projective theorem: Pappus of Alexandria proved that the three points (x, y, z) formed by intersecting the six lines that connect two sets of three collinear points (A, B, C; and D, E, F) are also collinear.

Let the distinct points A, B, C and D, E, F be on two different lines. Then the three intersection points—x of AE and BD, y of AF and CD, and z of BF and CE—are collinear. The second variant, by Pascal, as shown in the figure, uses certain properties of circles:

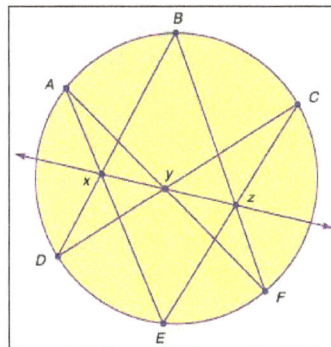

Pascal's projective theorem: The 17th-century French mathematician Blaise Pascal proved that the three points (x, y, z) formed by intersecting the six lines that connect any six distinct points (A, B, C, D, E, F) on a circle are collinear.

If the distinct points A, B, C, D, E, and F are on one circle, then the three intersection points x, y, and z (defined as above) are collinear. There is one more important invariant under projective mappings, known as the cross ratio. Given four distinct collinear points A, B, C, and D, the cross ratio is defined as:

$$CRat(A,B,C,D) = AC\,/\,BC \cdot BD\,/\,AD.$$

It may also be written as the quotient of two ratios:

$$CRat(A,B,C,D) = AC / BC : AD / BD.$$

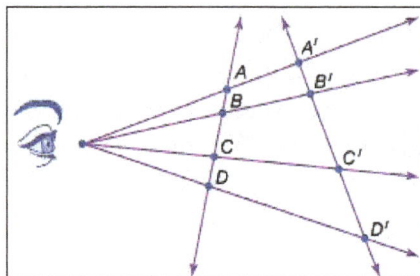

Cross ratio.

Although distances and ratios of distances are not preserved under projection, the cross ratio, defined as AC/BC · BD/AD, is preserved. That is, AC/BC · BD/AD = A′C′/B′C′ · B′D′/A′D′.

The latter formulation reveals the cross ratio as a ratio of ratios of distances. And while neither distance nor the ratio of distance is preserved under projection, Pappus first proved the startling fact that the cross ratio was invariant—that is,

$$CRat(A,B,C,D) = CRat(A',B',C',D').$$

However, this result remained a mere curiosity until its real significance became gradually clear in the 19th century as mappings became more and more important for transforming problems from one mathematical domain to another.

Projective Conic Sections

Conic sections can be regarded as plane sections of a right circular cone. By regarding a plane perpendicular to the cone's axis as the reality plane (RP), a "cutting" plane as the picture plane (PP), and the cone's apex as the projective "eye," each conic section can be seen to correspond to a projective image of a circle. Depending on the orientation of the cutting plane, the image of the circle will be a circle, an ellipse, a parabola, or a hyperbola.

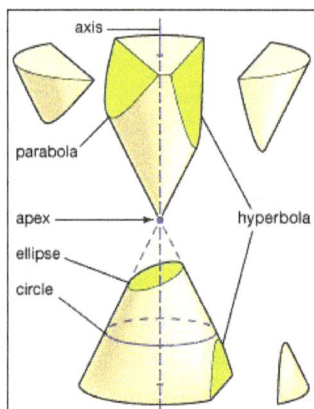

A plane passing through the apex and parallel to PP defines the line at infinity in the projective plane PP. The situation of Ω relative to RP determines the conic section in PP: If Ω intersects RP outside the base circle (the circle formed by the intersection of the cone and RP), the image of the circle will be an ellipse. If Ω is tangent to the base circle (in effect, tangent to the cone), the image will be a parabola. If Ω intersects the base circle (thus, cutting the circle in two), a hyperbola will result.

Pascal's theorem, quoted above, also follows easily for any conic section from its special case for the circle. Start by selecting six points on a conic section and project them back onto the base circle. As given earlier, the three relevant intersection points for six points on the circle will be collinear. Now project all nine points back to the conic section. Since collinear points (the three intersection points from the circle) are mapped onto collinear points, the theorem holds for any conic section. In this way the projective point of view unites the three different types of conics.

Similarly, more complicated curves and surfaces in higher-dimensional spaces can be unified through projections. For example, Isaac Newton showed that all plane curves defined by polynomials in x and y of degree 3 (the highest power of the variables is 3) can be obtained as projective images of just five types of polynomials.

References

- What-is-geometry, what-is-pure-math, pure-mathematics: uwaterloo.ca, Retrieved 13 June, 2019

- Euclidean-geometry, science: britannica.com, Retrieved 14 July, 2019

- Struik, D. J., A Source Book in Mathematics, 1200-1800, Harvard University Press, ISBN 978-0674823556

- Non-Euclidean-geometry, science britannica.com, Retrieved 15 August, 2019

- Differential-geometr, science: britannica.com, Retrieved 16 January, 2019

- Projective-geometry, science: britannica.com, Retrieved 17 February, 2019

Algebra 4

- **Elementary Algebra**
- **Abstract Algebra**
- **Universal Algebra**

Algebra is concerned with the study of mathematical symbols and the postulates used to manipulate these symbols. These symbols are used to represent numbers and quantities. It is divided into elementary algebra, abstract algebra, universal algebra, etc. This chapter closely examines these concepts of algebra to provide an extensive understanding of the subject.

Algebra is the branch of mathematics in which arithmetical operations and formal manipulations are applied to abstract symbols rather than specific numbers. The notion that there exists such a distinct subdiscipline of mathematics, as well as the term algebra to denote it, resulted from a slow historical development.

Classical Algebra

François Viète's work at the close of the 16th century, described in the section Viète and the formal equation, marks the start of the classical discipline of algebra. Further developments included several related trends, among which the following deserve special mention: the quest for systematic solutions of higher order equations, including approximation techniques; the rise of polynomials and their study as autonomous mathematical entities; and the increased adoption of the algebraic perspective in other mathematical disciplines, such as geometry, analysis, and logic. During this same period, new mathematical objects arose that eventually replaced polynomials as the main focus of algebraic study.

Analytic Geometry

The creation of what came to be known as analytic geometry can be attributed to two great 17th-century French thinkers: Pierre de Fermat and René Descartes. Using algebraic techniques developed by Viète and Girolamo Cardano, Fermat and Descartes tackled geometric problems that had remained unsolved since the time of the classical Greeks. The new kind of organic connection that they established between algebra and geometry was a major breakthrough, without which the subsequent development of mathematics in general, and geometry and calculus in particular, would be unthinkable.

In his famous book La Géométrie, Descartes established equivalences between algebraic operations and geometric constructions. In order to do so, he introduced a unit length that served as a reference for all other lengths and for all operations among them. For example, suppose that Descartes was given a segment AB and was asked to find its square root. He would draw the straight line DB, where DA was defined as the unit length. Then he would bisect DB at C, draw the semicircle on the diameter DB with centre C, and finally draw the perpendicular from A to E on the semicircle. Elementary properties of the circle imply that $\angle DEB = 90°$, which in turn implies that $\angle ADE = \angle AEB$ and $\angle DEA = \angle EBA$. Thus, $\triangle DEA$ is similar to $\triangle EBA$, or in other words, the ratio of corresponding sides is equal. Substituting x, 1, and y for AB, DA, and AE, respectively, one obtains $x/y = y/1$. Simplifying, $x = y^2$, or y is the square root of x. Thus, in what might appear to be an ordinary application of classical Greek techniques, Descartes demonstrated that he could find the square root of any given number, as represented by a line segment. The key step in his construction was the introduction of the unit length DA. This seemingly trivial move, or anything similar to it, had never been done before, and it had enormous repercussions for what could thereafter be done by applying algebraic reasoning to geometry.

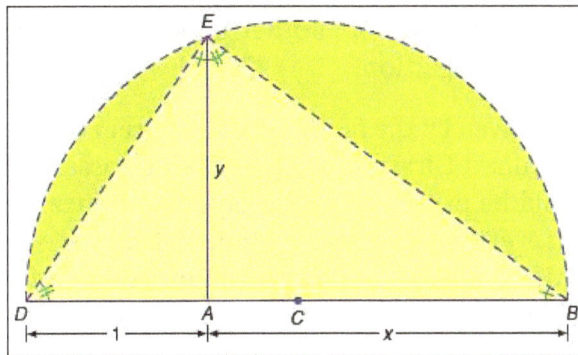

The French mathematician René Descartes demonstrated that the square root of any line segment could be constructed by the simple, but ingenious, addition of a line segment with unit length.

Descartes also introduced a notation that allowed great flexibility in symbolic manipulation. For instance, he would write,

$$\sqrt{C.a^3 - b^3 + abb}$$

to denote the cubic root of this algebraic expression. This was a direct continuation (with some improvement) of techniques and notations introduced by Viète. Descartes also introduced a new idea with truly far-reaching consequences when he explicitly eliminated the demand for homogeneity among the terms in an equation—although for convenience he tried to stick to homogeneity wherever possible.

Descartes's program was based on the idea that certain geometric loci (straight lines, circles, and conic sections) could be characterized in terms of specific kinds of equations involving magnitudes that were taken to represent line segments. However, he did not envision the equally important, reciprocal idea of finding the curve that corresponded to an arbitrary algebraic expression. Descartes was aware that much information about the properties of a curve—such as its tangents and enclosed areas—could be derived from its equation, but he did not elaborate.

On the other hand, Descartes was the first to discuss separately and systematically the algebraic properties of polynomial equations. This included his observations on the correspondence between the degree of an equation and the number of its roots, the factorization of a polynomial with known roots into linear factors, the rule for counting the number of positive and negative roots of an equation, and the method for obtaining a new equation whose roots were equal to those of a given equation, though increased or diminished by a given quantity.

The Fundamental Theorem of Algebra

Descartes's work was the start of the transformation of polynomials into an autonomous object of intrinsic mathematical interest. To a large extent, algebra became identified with the theory of polynomials. A clear notion of a polynomial equation, together with existing techniques for solving some of them, allowed coherent and systematic reformulations of many questions that had previously been dealt with in a haphazard fashion. High on the agenda remained the problem of finding general algebraic solutions for equations of degree higher than four. Closely related to this was the question of the kinds of numbers that should count as legitimate solutions, or roots, of equations. Attempts to deal with these two important problems forced mathematicians to realize the centrality of another pressing question, namely, the number of solutions for a given polynomial equation.

The answer to this question is given by the fundamental theorem of algebra, first suggested by the French-born mathematician Albert Girard in 1629, and which asserts that every polynomial with real number coefficients could be expressed as the product of linear and quadratic real number factors or, alternatively, that every polynomial equation of degree n with complex coefficients had n complex roots. For example, $x^3 + 2x^2 - x - 2$ can be decomposed into the quadratic factor $x^2 - 1$ and the linear factor $x + 2$, that is, $x^3 + 2x^2 - x - 2 = (x^2 - 1)(x + 2)$. The mathematical beauty of having n solutions for n-degree equations overcame most of the remaining reluctance to consider complex numbers as legitimate.

Although every single polynomial equation had been shown to satisfy the theorem, the essence of mathematics since the time of the ancient Greeks has been to establish universal principles. Therefore, leading mathematicians throughout the 18th century sought the honour of being the first to prove the theorem. The flaws in their proofs were generally related to the lack of rigorous foundations for polynomials and the various number systems. Indeed, the process of criticism and revision that accompanied successive attempts to formulate and prove some correct version of the theorem contributed to a deeper understanding of both.

The first complete proof of the theorem was given by the German mathematician Carl Friedrich Gauss in his doctoral dissertation of 1799. Subsequently, Gauss provided three additional proofs. A remarkable feature of all these proofs was that they were based on methods and ideas from calculus and geometry, rather than algebra. The theorem was fundamental in that it established the most basic concept around which the discipline as a whole was built. The theorem was also fundamental from the historical point of view, since it contributed to the consolidation of the discipline, its main tools, and its main concepts.

Impasse with Radical Methods

A major breakthrough in the algebraic solution of higher-degree equations was achieved by the Italian-French mathematician Joseph-Louis Lagrange in 1770. Rather than trying to find a general solution for quintic equations directly, Lagrange attempted to clarify first why all attempts to do so had failed by investigating the known solutions of third- and fourth-degree equations. In particular, he noticed how certain algebraic expressions connected with those solutions remained invariant when the coefficients of the equations were permuted (exchanged) with one another. Lagrange was certain that a deeper analysis of this invariance would provide the key to extending existing solutions to higher-degree equations.

Using ideas developed by Lagrange, in 1799 the Italian mathematician Paolo Ruffini was the first to assert the impossibility of obtaining a radical solution for general equations beyond the fourth degree. He adumbrated in his work the notion of a group of permutations of the roots of an equation and worked out some basic properties. Ruffini's proofs, however, contained several significant gaps.

Between 1796 and 1801, in the framework of his seminal number-theoretical investigations, Gauss systematically dealt with cyclotomic equations: $x^p - 1 = 0$ ($p > 2$ and prime). Although his new methods did not solve the general case, Gauss did demonstrate solutions for these particular higher-degree equations.

In 1824 the Norwegian mathematician Niels Henrik Abel provided the first valid proof of the impossibility of obtaining radical solutions for general equations beyond the fourth degree. However, this did not end polynomial research; rather, it opened an entirely new field of research since, as Gauss's example showed, some equations were indeed solvable. In 1828 Abel suggested two main points for research in this regard: to find all equations of a given degree solvable by radicals, and to decide if a given equation can be solved by radicals. His early death in complete poverty, two days before receiving an announcement that he had been appointed professor in Berlin, prevented Abel from undertaking this program.

Galois Theory

Rather than establishing whether specific equations can or cannot be solved by radicals, as Abel had suggested, the French mathematician Évariste Galois pursued the somewhat more general problem of defining necessary and sufficient conditions for the solvability of any given equation. Although Galois's life was short and exceptionally turbulent—he was arrested several times for supporting Republican causes, and he died the day before his 21st birthday from wounds incurred in a duel—his work reshaped the discipline of algebra.

Galois's Work on Permutations

Prominent among Galois's seminal ideas was the clear realization of how to formulate precise solvability conditions for a polynomial in terms of the properties of its group of permutations. A permutation of a set, say the elements a, b, and c, is any re-ordering of the elements, and it is usually denoted as follows:

$$\begin{pmatrix} a & b & c \\ c & a & b \end{pmatrix}$$

This particular permutation takes a to c, b to a, and c to b. For three elements, as here, there are six different possible permutations. In general, for n elements there are n! permutations to choose from. (*Where* $n! = n(n-1)(n-2)\cdots 2.1$). Furthermore, two permutations can be combined to produce a third permutation in an operation known as composition. (The set of permutations are closed under the operation of composition). For example,

$$\begin{pmatrix} a & b & c \\ c & a & b \end{pmatrix} * \begin{pmatrix} a & b & c \\ a & c & b \end{pmatrix} = \begin{pmatrix} a & b & c \\ b & a & c \end{pmatrix}$$

Here a goes first to c (in the first permutation) and then from c to b (in the second permutation), which is equivalent to a going directly to b, as given by the permutation to the right of the equation. Composition is associative—given three permutations P, Q, and R, then $(P*Q)*R = P*(Q*R)$. Also, there exists an identity permutation that leaves the elements unchanged:

$$I = \begin{pmatrix} a & b & c \\ a & b & c \end{pmatrix}$$

Finally, for each permutation there exists another permutation, known as its inverse, such that their composition results in the identity permutation. The set of permutations for n elements is known as the symmetric group S_n. The concept of an abstract group developed somewhat later. It consisted of a set of abstract elements with an operation defined on them such that the conditions given above were satisfied: closure, associativity, an identity element, and an inverse element for each element in the set. This abstract notion is not fully present in Galois's work. Like some of his predecessors, Galois focused on the permutation group of the roots of an equation. Through some beautiful and highly original mathematical ideas, Galois showed that a general polynomial equation was solvable by radicals if and only if its associated symmetric group was "soluble." Galois's result, it must be stressed, referred to conditions for a solution to exist; it did not provide a way to calculate radical solutions in those cases where they existed.

Acceptance of Galois Theory

Galois's work was both the culmination of a main line of algebra—solving equations by radical methods—and the beginning of a new line—the study of abstract structures. Work on permutations, started by Lagrange and Ruffini, received further impetus in 1815 from the leading French mathematician, Augustin-Louis Cauchy. In a later work of 1844, Cauchy systematized much of this knowledge and introduced basic concepts. For instance, the permutation,

$$\begin{pmatrix} a & b & c & d & e \\ b & a & e & c & d \end{pmatrix}$$

was denoted by Cauchy in cycle notation as (ab)(ced), meaning that the permutation was obtained by the disjoint cycles a to b (and back to a) and c to e to d (and back to c).

A series of unusual and unfortunate events involving the most important contemporary French mathematicians prevented Galois's ideas from being published for a long time. It was not until 1846 that Joseph Liouville edited and published for the first time, in his prestigious Journal de Mathématiques Pures et Appliquées, the important memoire in which Galois had presented his main ideas and that the Paris Academy had turned down in 1831. In Germany, Leopold Kronecker applied some of these ideas to number theory in 1853, and Richard Dedekind lectured on Galois theory in 1856. At this time, however, the impact of the theory was still minimal.

A major turning point came with the publication of Traité des substitutions et des équations alge-briques by the French mathematician Camille Jordan. In his book and papers, Jordan elaborated an abstract theory of permutation groups, with algebraic equations merely serving as an illustrative application of the theory. In particular, Jordan's treatise was the first group theory book and it served as the foundation for the conception of Galois theory as the study of the interconnections between extensions of fields and the related Galois groups of equations—a conception that proved fundamental for developing a completely new abstract approach to algebra in the 1920s. Major contributions to the development of this point of view for Galois theory came variously from Enrico Betti in Italy and from Dedekind, Henrich Weber, and Emil Artin in Germany.

Applications of Group Theory

Galois theory arose in direct connection with the study of polynomials, and thus the notion of a group developed from within the mainstream of classical algebra. However, it also found im-portant applications in other mathematical disciplines throughout the 19th century, particularly geometry and number theory.

Geometry

In 1872 Felix Klein suggested in his inaugural lecture at the University of Erlangen, Germany, that group theoretical ideas might be fruitfully put to use in the context of geometry. Since the beginning of the 19th century, the study of projective geometry had attained renewed impe-tus, and later on non-Euclidean geometries were introduced and increasingly investigated. This proliferation of geometries raised pressing questions concerning both the interrelations among them and their relationship with the empirical world. Klein suggested that these geometries could be classified and ordered within a conceptual hierarchy. For instance, projective geometry seemed particularly fundamental because its properties were also relevant in Euclidean geom-etry, while the main concepts of the latter, such as length and angle, had no significance in the former.

A geometric hierarchy may be expressed in terms of which transformations leave the most relevant properties of a particular geometry unchanged. It turned out that these sets of transformations were best understood as forming a group. Klein's idea was that the hierarchy of geometries might be reflected in a hierarchy of groups whose properties would be easier to understand. An example from Euclidean geometry illustrates the basic idea. The set of rotations in the plane has closure: if rotation I rotates a figure by an angle α, and rotation J by an angle β, then rotation $I*J$ ro-tates it by an angle $\alpha + \beta$. The rotation operation is obviously associative, $\alpha + (\beta + \gamma) = (\alpha + \beta) + \gamma$. The identity element is the rotation through an angle of 0 degrees, and the inverse of the rotation through angle α is the angle $-\alpha$. Thus the set of rotations of the plane is a group of invariant transformations for Euclidean geometry. The groups associated with other kinds of geometries is somewhat more involved, but the idea remains the same.

In the 1880s and '90s, Klein's friend, the Norwegian Sophus Lie, undertook the enormous task of classifying all possible continuous groups of geometric transformations, a task that eventual-ly evolved into the modern theory of Lie groups and Lie algebras. At roughly the same time, the French mathematician Henri Poincaré studied the groups of motions of rigid bodies, a work that helped to establish group theory as one of the main tools in modern geometry.

Number Theory

The notion of a group also started to appear prominently in number theory in the 19th century, especially in Gauss's work on modular arithmetic. In this context, he proved results that were later reformulated in the abstract theory of groups—for instance (in modern terms), that in a cyclic group (all elements generated by repeating the group operation on one element) there always exists a subgroup of every order (number of elements) dividing the order of the group.

In 1854 Arthur Cayley, one of the most prominent British mathematicians of his time, was the first explicitly to realize that a group could be defined abstractly—without any reference to the nature of its elements and only by specifying the properties of the operation defined on them. Generalizing on Galois's ideas, Cayley took a set of meaningless symbols 1, α, β,... with an operation defined on them as shown in the table below:

	1	α	β	...
1	1	α	β	...
α	α	α^2	$\alpha\beta$	...
β	β	$\beta\alpha$	β^2	...
...	...	...	...	...

Cayley demanded only that the operation be closed with respect to the elements on which it was defined, while he assumed implicitly that it was associative and that each element had an inverse. He correctly deduced some basic properties of the group, such as that if the group has n elements, then $\theta^n = 1$ for each element θ. Nevertheless, in 1854 the idea of permutation groups was rather new, and Cayley's work had little immediate impact.

Fundamental Concepts of Modern Algebra

Prime Factorization

Some other fundamental concepts of modern algebra also had their origin in 19th-century work on number theory, particularly in connection with attempts to generalize the theorem of (unique) prime factorization beyond the natural numbers. This theorem asserted that every natural number could be written as a product of its prime factors in a unique way, except perhaps for order (e.g., $24 = 2 \cdot 2 \cdot 2 \cdot 3$). This property of the natural numbers was known, at least implicitly, since the time of Euclid. In the 19th century, mathematicians sought to extend some version of this theorem to the complex numbers.

One should not be surprised, then, to find the name of Gauss in this context. In his classical investigations on arithmetic Gauss was led to the factorization properties of numbers of the type $a + ib$ (a and b integers and i = Square root of $i = \sqrt{-1}$), sometimes called Gaussian integers. In doing so, Gauss not only used complex numbers to solve a problem involving ordinary integers, a fact remarkable in itself, but he also opened the way to the detailed investigation of special subdomains of the complex numbers.

In 1832 Gauss proved that the Gaussian integers satisfied a generalized version of the factorization theorem where the prime factors had to be especially defined in this domain. In the 1840s the German mathematician Ernst Eduard Kummer extended these results to other, even more general domains of complex numbers, such as numbers of the form $a + \theta b$, where $\theta^2 = n$ for n a fixed integer, or numbers of the form $a + \rho b$, where $\rho^n = 1$, $\rho \neq 1$, and $n > 2$. Although Kummer did prove interesting results, it finally turned out that the prime factorization theorem was not valid in such general domains. The following example illustrates the problem.

Consider the domain of numbers of the form $a + b\sqrt{-5}$ and, in particular, the number $21 = 21 + 0\sqrt{-5}$. 21 can be factored as both $3\cdot7$ and as $(4+\sqrt{-5})(4-\sqrt{-5})$. It can be shown that none of the numbers $3, 7, 4\pm\sqrt{-5}$ could be further decomposed as a product of two different numbers in this domain. Thus, in one sense they were prime. However, at the same time they violated a property of prime numbers known from the time of Euclid: if a prime number p divides a product ab, then it either divides a or b. In this instance, 3 divides 21 but neither of the factors $4 + \sqrt{-5}$ or $4 - \sqrt{-5}$.

This situation led to the concept of indecomposable numbers. In classical arithmetic any indecomposable number is a prime (and vice versa), but in more general domains a number may be indecomposable, such as 3 here, yet not prime in the earlier sense. The question thus remained open which domains the prime factorization theorem was valid in and how properly to formulate a generalized version of it. This problem was undertaken by Dedekind in a series of works spanning over 30 years, starting in 1871. Dedekind's general methodological approach promoted the introduction of new concepts around which entire theories could be built. Specific problems were then solved as instances of the general theory.

Fields

A main question pursued by Dedekind was the precise identification of those subsets of the complex numbers for which some generalized version of the theorem made sense. The first step toward answering this question was the concept of a field, defined as any subset of the complex numbers that was closed under the four basic arithmetic operations (except division by zero). The largest of these fields was the whole system of complex numbers, whereas the smallest field was the rational numbers. Using the concept of field and some other derivative ideas, Dedekind identified the precise subset of the complex numbers for which the theorem could be extended. He named that subset the algebraic integers.

Ideals

Finally, Dedekind introduced the concept of an ideal. A main methodological trait of Dedekind's innovative approach to algebra was to translate ordinary arithmetic properties into properties of sets of numbers. In this case, he focused on the set I of multiples of any given integer and pointed out two of its main properties:

- If n and m are two numbers in I, then their difference is also in I.

- If n is a number in I and a is any integer, then their product is also in I.

As he did in many other contexts, Dedekind took these properties and turned them into definitions. He defined a collection of algebraic integers that satisfied these properties as an ideal in the complex numbers. This was the concept that allowed him to generalize the prime factorization theorem in distinctly set-theoretical terms.

In ordinary arithmetic, the ideal generated by the product of two numbers equals the intersection of the ideals generated by each of them. For instance, the set of multiples of 6 (the ideal generated by 6) is the intersection of the ideal generated by 2 and the ideal generated by 3. Dedekind's generalized versions of the theorem were phrased precisely in these terms for general fields of complex numbers and their related ideals. He distinguished among different types of ideals and different types of decompositions, but the generalizations were all-inclusive and precise. More important, he reformulated what were originally results on numbers, their factors, and their products as far more general and abstract results on special domains, special subsets of numbers, and their intersections.

Dedekind's results were important not only for a deeper understanding of factorization. He also introduced the set-theoretical approach into algebraic research, and he defined some of the most basic concepts of modern algebra that became the main focus of algebraic research throughout the 20th century. Moreover, Dedekind's ideal-theoretical approach was soon successfully applied to the factorization of polynomials as well, thus connecting itself once again to the main focus of classical algebra.

Systems of Equations

In spite of the many novel algebraic ideas that arose in the 19th century, solving equations and studying properties of polynomial forms continued to be the main focus of algebra. The study of systems of equations led to the notion of a determinant and to matrix theory.

Determinants

Given a system of n linear equations in n unknowns, its determinant was defined as the result of a certain combination of multiplication and addition of the coefficients of the equations that allowed the values of the unknowns to be calculated directly. For example, given the system,

$$a_1 x + b_1 y = c_1$$

$$a_2 x + b_2 y = c_2$$

the determinant Δ of the system is the number $\Delta = a_1 b_2 - a_2 b_1$, and the values of the unknowns are given by,

$$x = (c_1 b_2 - c_2 b_1) / \Delta$$

$$y = (a_1 c_2 - a_2 c_1) / \Delta.$$

Historians agree that the 17th-century Japanese mathematician Seki Kōwa was the earliest to use methods of this kind systematically. In Europe, credit is usually given to his contemporary, the German coinventor of calculus, Gottfried Wilhelm Leibniz.

In 1815 Cauchy published the first truly systematic and comprehensive study of determinants, and he was the one who coined the name. He introduced the notation $(a_{i,n})$ for the system of coefficients of the system and demonstrated a general method for calculating the determinant.

Matrices

Closely related to the concept of a determinant was the idea of a matrix as an arrangement of numbers in lines and columns. That such an arrangement could be taken as an autonomous mathematical object, subject to special rules that allow for manipulation like ordinary numbers, was first conceived in the 1850s by Cayley and his good friend the attorney and mathematician James Joseph Sylvester. Determinants were a main, direct source for this idea, but so were ideas contained in previous work on number theory by Gauss and by the German mathematician Ferdinand Gotthold Max Eisenstein.

Given a system of linear equations:

$$\xi = \alpha x + \beta y + \gamma z + \ldots$$
$$\eta = \alpha' x + \beta' y + \gamma' z + \ldots$$
$$\zeta = \alpha'' x + \beta'' y + \gamma'' z + \ldots$$
$$\ldots = \ldots + \ldots + \ldots + \ldots$$

Cayley represented it with a matrix as follows:

$$(\xi, \eta, \zeta, \ldots) = \begin{pmatrix} \alpha & \beta & \gamma & \ldots \\ \alpha' & \beta' & \gamma' & \ldots \\ \alpha'' & \beta'' & \gamma'' & \ldots \\ \ldots & \ldots & \ldots & \ldots \end{pmatrix} (x, y, z, \ldots)$$

The solution could then be written as:

$$(x, y, z, \ldots) = \begin{pmatrix} \alpha & \beta & \gamma & \ldots \\ \alpha' & \beta' & \gamma' & \ldots \\ \alpha'' & \beta'' & \gamma'' & \ldots \\ \ldots & \ldots & \ldots & \ldots \end{pmatrix}^{-1} (\xi, \eta, \zeta, \ldots)$$

The matrix bearing the −1 exponent was called the inverse matrix, and it held the key to solving the original system of equations. Cayley showed how to obtain the inverse matrix using the determinant of the original matrix. Once this matrix is calculated, the arithmetic of matrices allowed him to solve the system of equations by a simple analogy with linear equations: $ax = b \rightarrow x = a^{-1}b$.

Cayley was joined by other mathematicians, such as the Irish William Rowan Hamilton, the German Georg Frobenius, and Jordan, in developing the theory of matrices, which soon became a fundamental tool in analysis, geometry, and especially in the emerging discipline of linear algebra. A further important point was that matrices enlarged the range of algebraic notions. In particular, matrices embodied a new, mathematically significant instance of a system with a well-elaborated arithmetic, whose rules departed from traditional number systems in the important sense that multiplication was not generally commutative.

In fact, matrix theory was naturally connected after 1830 with a central trend in British mathematics developed by George Peacock and Augustus De Morgan, among others. In trying to overcome the last reservations about the legitimacy of the negative and complex numbers, these mathematicians suggested that algebra be conceived as a purely formal, symbolic language, irrespective of the nature of the objects whose laws of combination it stipulated. In principle, this view allowed for new, different kinds of arithmetic, such as matrix arithmetic. The British tradition of symbolic algebra was instrumental in shifting the focus of algebra from the direct study of objects (numbers, polynomials, and the like) to the study of operations among abstract objects. Still, in most respects, Peacock and De Morgan strove to gain a deeper understanding of the objects of classical algebra rather than to launch a new discipline.

Another important development in Britain concerned the elaboration of an algebra of logic. De Morgan and George Boole, and somewhat later Ernst Schröder in Germany, were instrumental in transforming logic from a purely metaphysical into a mathematical discipline. They also added to the growing realization of the immense potential of algebraic thinking, freed from its narrow conception as the discipline of polynomial equations and number systems.

Quaternions and Vectors

Remaining doubts about the legitimacy of complex numbers were finally dispelled when their geometric interpretation became widespread among mathematicians. This interpretation, initially and independently conceived by the Norwegian surveyor Caspar Wessel and the French bookkeeper Jean-Robert Argand was made known to a larger audience of mathematicians mainly through its explicit use by Gauss in his 1848 proof of the fundamental theorem of algebra. Under this interpretation, every complex number appeared as a directed segment on the plane, characterized by its length and its angle of inclination with respect to the x-axis. The number i thus corresponded to the segment of length 1 that was perpendicular to the x-axis. Once a proper arithmetic was defined on these numbers, it turned out that $i^2 = -1$, as expected.

An alternative interpretation, very much within the spirit of the British school of symbolic algebra, was published in 1837 by Hamilton. Hamilton defined a complex number $a+bi$ as a pair (a,b) of real numbers and gave the laws of arithmetic for such pairs. For example, he defined multiplication as:

$$(a,b)(c,d) = (ac-bd, bc+ad).$$

In Hamilton's notation $i = (0, 1)$ and by the above definition of complex multiplication $(0, 1)(0, 1) = (-1, 0)$ — that is, $i^2 = -1$ as desired. This formal interpretation obviated the need to give any essentialist definition of complex numbers.

Starting in 1830, Hamilton pursued intensely, and unsuccessfully, a scheme to extend his idea to triplets (a, b, c), which he expected to be of great utility in mathematical physics. His difficulty lay in defining a consistent multiplication for such a system, which in hindsight is known to be impossible. In 1843 Hamilton finally realized that the generalization he was looking for had to be found in the system of quadruplets (a, b, c, d), which he named quaternions. He wrote them, in analogy with the complex numbers, as $a+bi+cj+dk$, and his new arithmetic was based on the rules: $i^2 = j^2 = k^2 = ijk = -1$ and $ij = k, ji = -k, jk = i, kj = -i, ki = j$, and $ik = -j$. This was the first example of a coherent, significant mathematical system that preserved all of the laws of ordinary arithmetic, with the exception of commutativity.

In spite of Hamilton's initial hopes, quaternions never really caught on among physicists, who generally preferred vector notation when it was introduced later. Nevertheless, his ideas had an enormous influence on the gradual introduction and use of vectors in physics. Hamilton used the name scalar for the real part a of the quaternion, and the term vector for the imaginary part $bi + cj + dk$, and defined what are now known as the scalar (or dot) and vector (or cross) products. It was through successive work in the 19th century of the Britons Peter Guthrie Tait, James Clerk Maxwell, and Oliver Heaviside and the American Josiah Willard Gibbs that an autonomous theory of vectors was first established while developing on Hamilton's initial ideas. In spite of physicists' general lack of interest in quaternions, they remained important inside mathematics, although mainly as an example of an alternate algebraic system.

The Close of the Classical Age

The last major algebra textbook in the classical tradition was Heinrich Weber's Lehrbuch der Algebra, which codified the achievements and current dominant views of the subject and remained highly influential for several decades. At its centre was a well-elaborated, systematic conception of the various systems of numbers, built as a rigorous hierarchy from the natural numbers up to the complex numbers. Its primary focus was the study of polynomials, polynomial equations, and polynomial forms, and all relevant results and methods derived in the book directly depended on the properties of the systems of numbers. Radical methods for solving equations received a great deal of attention, but so did approximation methods, which are now typically covered instead in analysis and numerical analysis textbooks. Recently developed concepts, such as groups and fields, as well as methods derived from Galois's work, were treated in Weber's textbook, but only as useful tools to help deal with the main topic of polynomial equations.

To a large extent, Weber's textbook was a very fine culmination of a long process that started in antiquity. Fortunately, rather than bring this process to a conclusion, it served as a catalyst for the next stage of algebra.

Structural Algebra

At the turn of the 20th century, algebra reflected a very clear conceptual hierarchy based on a systematically elaborated arithmetic, with a theory of polynomial equations built on top of it. Finally, a well-developed set of conceptual tools, most prominently the idea of groups, offered a comprehensive means of investigating algebraic properties. Then in 1930 a textbook was published that presented a totally new image of the discipline. This was Moderne Algebra, by the Dutch mathematician Bartel van der Waerden, who since 1924 had attended lectures in Germany by Emmy Noether at Göttingen and by Emil Artin at Hamburg. Van der Waerden's new image of the discipline inverted the conceptual hierarchy of classical algebra. Groups, fields, rings, and other related concepts became the main focus, based on the implicit realization that all of these concepts were, in fact, instances of a more general, underlying idea: the idea of an algebraic structure. Thus, the main task of algebra became the elucidation of the properties of each of these structures and of the relationships among them. Similar questions were now asked about all these concepts, and similar concepts and techniques were used where possible. The main tasks of classical algebra became ancillary. The systems of real numbers, rational numbers, and polynomials were studied as particular instances of certain algebraic structures; the properties of these systems depended on what was known about the general structures of which they were instances, rather than the other way round.

Precursors to the Structural Approach

Van der Waerden's book did not contain many new results or concepts. Its innovation lay in the unified picture it presented of the discipline of algebra. Van der Waerden brought together, in a surprisingly illuminating manner, algebraic research that had taken place over the previous three decades and in doing so he combined the contributions of several leading German algebraists from the beginning of the 20th century.

Hilbert and Steinitz

Of these German mathematicians, few were more important than David Hilbert. Among his important contributions, his work in the 1890s on the theory of algebraic number fields was decisive in establishing the conceptual approach promoted by Dedekind as dominant for several decades. As the undisputed leader of mathematics at Göttingen, then the world's premiere research institution, Hilbert's influence propagated through the 68 doctoral dissertations he directed as well as through the many students and mathematicians who attended his lectures. To a significant extent, the structural view of algebra was the product of some of Hilbert's innovations, yet he basically remained a representative of the classical discipline of algebra. It is likely that the kind of algebra that developed under the influence of van der Waerden's book had no direct appeal for Hilbert.

In 1910 Ernst Steinitz published an influential article on the abstract theory of fields that was an important milestone on the road to the structural image of algebra. His work was highly structural in that he first established the simplest kinds of subfields that any field contains and established a classification system. He then investigated how properties were passed from a field to any extension of it or to any of its subfields. In this way, he was able to characterize all possible fields abstractly. To a great extent, van der Waerden extended to the whole discipline of algebra what Steinitz accomplished for the more restricted domain of fields.

Noether and Artin

The greatest influence behind the consolidation of the structural image of algebra was no doubt Noether, who became the most prominent figure in Göttingen in the 1920s. Noether synthesized the ideas of Dedekind, Hilbert, Steinitz, and others in a series of articles in which the theory of factorization of algebraic numbers and of polynomials was masterly and succinctly subsumed under a single theory of abstract rings. She also contributed important papers to the theory of hypercomplex systems (extensions, such as the quaternions, of complex numbers to higher dimensions) that followed a similar approach, further demonstrating the potential of the structural approach.

The last significant influence on van der Waerden's structural image of algebra was by Artin, above all for the latter's reformulation of Galois theory. Rather than speaking of the Galois group of a polynomial equation with coefficients in a particular field, Artin focused on the group of automorphisms of the coefficients' splitting field (the smallest extension of the field such that the polynomial could be factored into linear terms). Galois theory could then be seen as the study of the interrelations between the extensions of a field and the possible subgroups of the Galois group of the original field. In this typical structural reformulation of a classical 19th-century theory of algebra, the problem of solvability of equations by radicals appeared as a particular application of an abstract general theory.

The Structural Approach Dominates

After the late 1930s it was clear that algebra, and in particular the structural approach within it, had become one of the most dynamic areas of research in mathematics. Structural methods, results, and concepts were actively pursued by algebraists in Germany, France, the United States, Japan, and elsewhere. The structural approach was also successfully applied to re-define other mathematical disciplines. An important early example of this was the thorough reformulation of algebraic geometry in the hands of van der Waerden, André Weil in France, and the Russian-born Oscar Zariski in Italy and the United States. In particular, they used the concepts and approach developed in ring theory by Noether and her successors. Another important example was the work of the American Marshall Stone, who in the late 1930s defined Boolean algebras, bringing under a purely algebraic framework ideas stemming from logic, topology, and algebra itself.

Over the following decades, algebra textbooks appeared around the world along the lines established by van der Waerden. Prominent among these was A Survey of Modern Algebra by Saunders Mac Lane and Garret Birkhoff, a book that was fundamental for the next several generations of mathematicians in the United States. Nevertheless, it must be stressed that not all algebraists felt, at least initially, that the new direction implied by Moderne Algebra was paramount. More classically oriented research was still being carried out well beyond the 1930s. The research of Frobenius and his former student Issai Schur, who were the most outstanding representatives of the Berlin mathematical school at the beginning of the 20th century, and of Hermann Weyl, one of Hilbert's most prominent students, merit special mention.

Algebraic Superstructures

Although the structural approach had become prominent in many mathematical disciplines, the notion of structure remained more a regulative, informal principle than a real mathematical concept for independent investigation. It was only natural that sooner or later the question would arise how to define structures in such a way that the concept could be investigated. For example, Noether brought new and important insights into certain rings (algebraic numbers and polynomials) previously investigated under separate frameworks by studying their underlying structures. Similarly, it was expected that a general metatheory of structures, or superstructures, would prove fruitful for studying other related concepts.

Bourbaki

Attempts to develop such a metatheory were undertaken starting in the 1940s. The first one came from a group of young French mathematicians working under the common pseudonym of Nicolas Bourbaki. The founders of the group included Weil, Jean Dieudonné, and Henri Cartan. Over the next few decades, the group published a collection of extremely influential textbooks, Eléments de mathématique, that covered several central mathematical disciplines, particularly from a structural perspective. Yet, to the extent that Bourbaki's mathematics was structural, it was so in a general, informal way. As van der Waerden extended to all of algebra the structural approach that Steinitz introduced in the theory of fields, so Bourbaki's Eléments extended this approach to a truly broad range of mathematical disciplines. Although Bourbaki did define a formal concept of structure in the first book of the collection, their concept turned out to be quite cumbersome and was not pursued further.

Category Theory

The second attempt to formalize the notion of structure developed within category theory. The first paper on the subject was published in the United States in 1942 by Mac Lane and Samuel Eilenberg. The idea behind their approach was that the essential features of any particular mathematical domain (a category) could be identified by focusing on the interrelations among its elements, rather than looking at the behaviour of each element in isolation. For example, what characterized the category of groups were the properties of its homomorphisms (mappings between groups that preserve algebraic operations) and comparisons with morphisms for other categories, such as homeomorphisms for topological spaces. Another important concept of Mac Lane and Eilenberg was their formulation of "functors," a generalization of the idea of function that enabled them to connect different categories. For example, in algebraic topology functors associated topological spaces with certain groups such that their topological properties could be expressed as algebraic properties of the groups—a process that enabled powerful algebraic tools to be used on previously intractable problems.

Although category theory did not become a universal language for all of mathematics, it did become the standard formulation for algebraic topology and homology. Category theory also led to new approaches in the study of the foundations of mathematics by means of Topos theory. Some of these developments were further enhanced between 1956 and 1970 through the intensive work of Alexandre Grothendieck and his collaborators in France, using still more general concepts based on categories.

New Challenges and Perspectives

The enormous productivity of research in algebra over the second half of the 20th century precludes any complete synopsis. Nevertheless, two main issues deserve some comment. The first was a trend toward abstraction and generalization as embodied in the structural approach. This trend was not exclusive, however. Researchers moved back and forth, studying general structures as well as classical entities such as the real and rational numbers. The second issue was the introduction of new kinds of proofs and techniques. The following examples are illustrative.

A subgroup H of a group G is called a normal group if for every element g in G and h in H, g^{-1}hg is an element of H. A group with no normal subgroups is known as a simple group. Simple groups are the basic components of group theory, and since Galois's time it was known that the general quintic was unsolvable by radicals because its Galois group was simple. However, a full characterization of simple groups remained unattainable until a major breakthrough in 1963 by two Americans, Walter Feit and John G. Thomson, who proved an old conjecture of the British mathematician William Burnside, namely, that the order of noncommutative finite simple groups is always even. Their proof was long and involved, but it reinforced the belief that a full classification of finite simple groups might, after all, be possible. The completion of the task was announced in 1983 by the American mathematician Daniel Gorenstein, following the contributions of hundreds of individuals over thousands of pages. Although this classification seems comprehensive, it is anything but clear-cut and systematic, since simple groups appear in all kinds of situations and under many guises. Thus, there seems to be no single individual who can boast of knowing all of its details. This kind of very large, collective theorem is certainly a novel mathematical phenomenon.

Another example concerns the complex and involved question of the use of computers in proving and even formulating new theorems. This now incipient trend will certainly receive increased attention in the 21st century.

Finally, probabilistic methods of proof in algebra, and in particular for solving difficult, open problems in group theory, have been introduced. This trend began with a series of papers by the Hungarian mathematicians Paul Erdős and Paul Turán, both of whom introduced probabilistic methods into many other branches of mathematics as well.

Elementary Algebra

Elemetary algebra is the branch of mathematics that deals with the general properties of numbers and the relations between them. Algebra is fundamental not only to all further mathematics and statistics but to the natural sciences, computer science, economics, and business. Along with writing, it is a cornerstone of modern scientific and technological civilization. Earlier civilizations—Babylonian, Greek, Indian, Chinese, and Islamic—all contributed in important ways to the development of elementary algebra. It was left for Renaissance Europe, though, to develop an efficient system for representing all real numbers and a symbolism for representing unknowns, relations between them, and operations.

Elementary algebra is concerned with the following topics:

- Real and complex numbers, constants, and variables—collectively known as algebraic quantities.
- Rules of operation for such quantities.
- Geometric representations of such quantities.
- Formation of expressions involving algebraic quantities.
- Rules for manipulating such expressions.
- Formation of sentences, also called equations, involving algebraic expressions.
- Solution of equations and systems of equations.

Algebraic Quantities

The principal distinguishing characteristic of algebra is the use of simple symbols to represent numerical quantities and mathematical operations. Following a system that originated with the 17th-century French thinker René Descartes, letters near the beginning of the alphabet (a, b, c,...) typically represent known, but arbitrary, numbers in a problem, while letters near the end of the alphabet, especially x, y, and z, represent unknown quantities, or variables. The + and − signs indicate addition and subtraction of these quantities, but multiplication is simply indicated by adjacent letters. Thus, ax represents the product of a by x. This simple expression can be interpreted, for example, as the interest earned in one year by a sum of a dollars invested at an annual rate of x. It can also be interpreted as the distance traveled in a hours by a car moving at x miles per hour. Such flexibility of representation is what gives algebra its great utility.

Another feature that has greatly increased the range of algebraic applications is the geometric representation of algebraic quantities. For instance, to represent the real numbers, a straight line is imagined that is infinite in both directions. An arbitrary point O can be chosen as the origin, representing the number 0, and another arbitrary point U chosen to the right of O. The segment OU (or the point U) then represents the unit length, or the number 1. The rest of the positive numbers correspond to multiples of this unit length—so that 2, for example, is represented by a segment OV, twice as long as OU and extended in the same direction. Similarly, the negative real numbers extend to the left of O. A straight line whose points are thus identified with the real numbers is called a number line. Many earlier mathematicians realized there was a relationship between all points on a straight line and all real numbers, but it was the German mathematician Richard Dedekind who made this explicit as a postulate in his Continuity and Irrational Numbers.

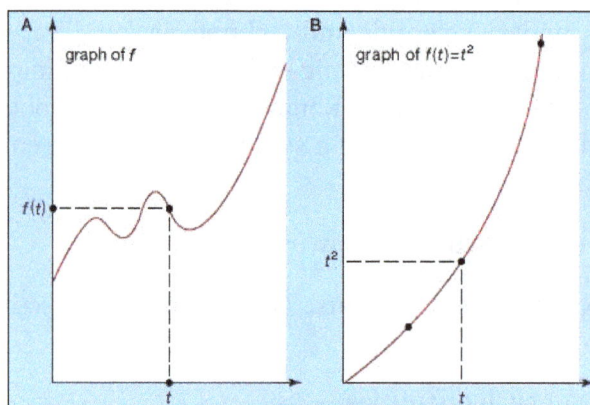

Graph of a function.

Part A illustrates the general idea of graphing any function: choose a value for the independent variable, t, calculate the corresponding value for f(t), and repeat this process until the general shape of the graph is apparent. (In practice, various techniques are available to reduce the number of values needed to determine the graph's basic shape). In part B a specific function, the parabola $f(t) = t^2$, is graphed for further illustration.

In the Cartesian coordinate system (named for Descartes) of analytic geometry, one horizontal number line (usually called the x-axis) and one vertical number line (the y-axis) intersect at right angles at their common origin to provide coordinates for each point in the plane. For example, the point on a vertical line through some particular x on the x-axis and on the horizontal line through some y on the y-axis is represented by the pair of real numbers (x, y). A similar geometric representation exists for the complex numbers, where the horizontal axis corresponds to the real numbers and the vertical axis corresponds to the imaginary numbers (where the imaginary unit i is equal to the square root of −1). The algebraic form of complex numbers is $x+iy$, where x represents the real part and iy the imaginary part.

This pairing of space and number gives a means of pairing algebraic expressions, or functions, in a single variable with geometric objects in the plane, such as straight lines and circles. The result of this pairing may be thought of as the graph of the expression for different values of the variable.

Algebraic Expressions

Any of the quantities mentioned so far may be combined in expressions according to the usual arithmetic operations of addition, subtraction, and multiplication. Thus, $ax+by$ and $axx+bx+c$ are common algebraic expressions. However, exponential notation is commonly used to avoid repeating the same term in a product, so that one writes x^2 for xx and y^3 for yyy. (By convention $x^0=1$.) Expressions built up in this way from the real and complex numbers, the algebraic quantities $a,b,c, ...,x,y,z$, and the three above operations are called polynomials—a word introduced in the late 16th century by the French mathematician François Viète from the Greek polys ("many") and the Latin nominem ("name" or "term"). One way of characterizing a polynomial is by the number of different unknown, or variable, quantities in it. Another way of characterizing a polynomial is by its degree. The degree of a polynomial in one unknown is the largest power of the unknown appearing in it. The expressions $ax+b, ax^2+bx+c$, and ax^3+bx^2+cx+d are general polynomials in one unknown (x) of degrees 1, 2, and 3, respectively. When only one unknown is involved, it does not matter which letter is used for it. One could equally well write the above polynomials as $ay+b, az^2+bz+c$, and at^3+bt^2+ct+d.

Because some insight into complicated functions can be obtained by approximating them with simpler functions, polynomials of the first degree were investigated early on. In particular, $ax+by=c$, which represents a straight line, and $ax+by+cz=e$, which represents a plane in three-dimensional space, were among the first algebraic equations studied.

Polynomials can be combined according to the three arithmetic operations of addition, subtraction, and multiplication, and the result is again a polynomial. To simplify expressions obtained by combining polynomials in this way, one uses the distributive law, as well as the commutative and associative laws for addition and multiplication. Until very recently a major drawback of algebra was the extreme tedium of routine manipulation of polynomials, but now a number of symbolic algebra programs make this work as easy as typing the expressions into a computer.

Common arithmetic properties

associative laws	$a+(b+c)=(a+b)+c$ and $a(bc)=(ab)c$
commutative laws	$a+b=b+a$ and $ab=ba$
distributive law	$a(b+c)=ab+ac$
additive identity	$0+a=a$
multiplicative identity	$1a=a$
additive inverse	$a+(-a)=0$
multiplicative inverse	$a(1/a)=1$, where $a\neq0$

By extending the operations on polynomials to include division, or ratios of polynomials, one obtains the rational functions. Examples of such rational functions are $2/3x$ and $(a+bx^2)/(c+dx^2+ex^5)$. Working with rational functions allows one to introduce the expression $1/x$ and its powers, $1/x^2$, $1/x^3$, ... (often written x^{-1}, x^{-2}, x^{-3}, ...). When the degree of the numerator of a rational function is at least as large as that of its denominator, it is possible to divide the numerator by the denominator much as one divides one integer by another. In this way one can write any rational function as the sum of a polynomial and a rational function in which the degree of the numerator is less than that of the denominator. For example,

$$(x^8 - x^5 + 3x^3 + 2)/(x^3 - 1) = x^5 + 3 + 5/(x^3 - 1).$$

Since this process reduces the degrees of the terms involved, it is especially useful for calculating the values of rational functions and for dealing with them when they arise in calculus.

Solving Algebraic Equations

For theoretical work and applications one often needs to find numbers that, when substituted for the unknown, make a certain polynomial equal to zero. Such a number is called a "root" of the polynomial. For example, the polynomial

$$-16t^2 + 88t + 48$$

represents the height above Earth at t seconds of a projectile thrown straight up at 88 feet per second from the top of a tower 48 feet high. (The 16 in the formula comes from one-half the acceleration of gravity, 32 feet per second per second.) By setting the equation equal to zero and factoring it as $(4t - 24)(-4t - 2) = 0$, the equation's one positive root is found to be 6, meaning that the object will hit the ground about 6 seconds after it is thrown. (This problem also illustrates the important algebraic concept of the zero factor property: if $ab = 0$, then either $a = 0$ or $b = 0$).

The theorem that every polynomial has as many complex roots as its degree is known as the fundamental theorem of algebra and was first proved in 1799 by the German mathematician Carl Friedrich Gauss. Simple formulas exist for finding the roots of the general polynomials of degrees one and two, and much less simple formulas exist for polynomials of degrees three and four. The French mathematician Évariste Galois discovered, shortly before his death in 1832, that no such formula exists for a general polynomial of degree greater than four. Many ways exist, however, of approximating the roots of these polynomials.

Solving Systems of Algebraic Equations

An extension of the study of single equations involves multiple equations that are solved simultaneously—so-called systems of equations. For example, the intersection of two straight lines, $ax + by = c$ and $Ax + By = C$, can be found algebraically by discovering the values of x and y that simultaneously solve each equation. The earliest systematic development of methods for solving systems of equations occurred in ancient China. An adaptation of a problem from the 1st-century-AD Chinese classic Nine Chapters on the Mathematical Procedures illustrates how such systems arise. Imagine there are two kinds of wheat and that you have four sheaves of the first type and five sheaves of the second type. Although neither of these is enough to produce a bushel of wheat, you can produce a bushel by adding three sheaves of the first type to five of the second type, or you can produce a bushel by adding four sheaves of the first type to two of the second type. What fraction of a bushel of wheat does a sheaf of each type of wheat contain?

Using modern notation, suppose we have two types of wheat, respectively, and x and y represent the number of bushels obtained per sheaf of the first and second types, respectively. Then the problem leads to the system of equations:

$$3x + 5y = 1 \; (bushel)$$

$$4x + 2y = 1 \; (bushel)$$

A simple method for solving such a system is first to solve either equation for one of the variables. For example, solving the second equation for y yields $y = 1/2 - 2x$. The right side of this equation can then be substituted for y in the first equation $(3x + 5y = 1)$, and then the first equation can be solved to obtain $x(=3/14)$. Finally, this value of x can be substituted into one of the earlier equations to obtain y $(=1/14)$. Thus, the first type yields $3/14$ bushels per sheaf and the second type yields $1/14$. Note that the solution $(3/14, 1/14)$ would be difficult to discern by graphing techniques. In fact, any precise value based on a graphing solution may be only approximate; for example, the point (0.0000001, 0) might look like (0, 0) on a graph, but even such a small difference could have drastic consequences in the real world.

Rather than individually solving each possible system of two equations in two unknowns, the general system can be solved. To return to the general equations given above:

$$ax + by = c$$

$$Ax + By = C$$

The solutions are given by $x = (Bc - bC)/(aB - Ab)$ and $y = (Ca - cA)/(aB - Ab)$. Note that the denominator of each solution, $(aB - Ab)$, is the same. It is called the determinant of the system, and systems in which the denominator is equal to zero have either no solution (in which case the equations represent parallel lines) or infinitely many solutions (in which case the equations represent the same line).

One can generalize simultaneous systems to consider m equations in n unknowns. In this case, one usually uses subscripted letters $x_1, x_2, ..., x_n$ for the unknowns and $a_{1,1}, ..., a_{1,n}; a_{2,1}, ..., a_{2,n}; ...; a_{m,1}, ..., a_{m,n}$, n for the coefficients of each equation, respectively. When $n = 3$ one is dealing with planes in three-dimensional space, and for higher values of n one is dealing with hyperplanes in spaces of higher dimension. In general, n equations in m unknowns have infinitely many solutions when $m < n$ and no solutions when $m > n$. The case $m = n$ is the only case where there can exist a unique solution.

Large systems of equations are generally handled with matrices, especially as implemented on computers.

Abstract Algebra

In algebra, which is a broad division of mathematics, abstract algebra (occasionally called modern algebra) is the study of algebraic structures. Algebraic structures include groups, rings, fields, modules, vector spaces, lattices, and algebras. The term *abstract algebra* was coined in the early 20th century to distinguish this area of study from the other parts of algebra.

Algebraic structures, with their associated homomorphisms, form mathematical categories. Category theory is a formalism that allows a unified way for expressing properties and constructions that are similar for various structures.

Universal algebra is a related subject that studies types of algebraic structures as single objects. For example, the structure of groups is a single object in universal algebra, which is called *variety of groups*.

The permutations of Rubik's Cube form a group, a fundamental concept within abstract algebra.

As in other parts of mathematics, concrete problems and examples have played important roles in the development of abstract algebra. Through the end of the nineteenth century, many – perhaps most – of these problems were in some way related to the theory of algebraic equations. Major themes include:

- Solving of systems of linear equations, which led to linear algebra.

- Attempts to find formulas for solutions of general polynomial equations of higher degree that resulted in discovery of groups as abstract manifestations of symmetry.

- Arithmetical investigations of quadratic and higher degree forms and diophantine equations, that directly produced the notions of a ring and ideal.

Numerous textbooks in abstract algebra start with axiomatic definitions of various algebraic structures and then proceed to establish their properties. This creates a false impression that in algebra axioms had come first and then served as a motivation and as a basis of further study. The true order of historical development was almost exactly the opposite. For example, the hypercomplex numbers of the nineteenth century had kinematic and physical motivations but challenged comprehension. Most theories that are now recognized as parts of algebra started as collections of disparate facts from various branches of mathematics, acquired a common theme that served as a core around which various results were grouped, and finally became unified on a basis of a common set of concepts. An archetypical example of this progressive synthesis can be seen in the history of group theory.

Early Group Theory

There were several threads in the early development of group theory, in modern language loosely corresponding to number theory, theory of equations, and geometry.

Leonhard Euler considered algebraic operations on numbers modulo an integer, modular arithmetic, in his generalization of Fermat's little theorem. These investigations were taken much further by Carl Friedrich Gauss, who considered the structure of multiplicative groups of residues mod n and established many properties of cyclic and more general abelian groups that arise in this way. In his investigations of composition of binary quadratic forms, Gauss explicitly stated the associative law for the composition of forms, but like Euler before him, he seems to have been more interested in concrete results than in general theory. In 1870, Leopold Kronecker gave a definition of an abelian group in the context of ideal class groups of a number field, generalizing Gauss's work; but it appears he did not tie his definition with previous work on groups, particularly permutation groups. In 1882, considering the same question, Heinrich M. Weber realized the connection and gave a similar definition that involved the cancellation property but omitted the existence of the inverse element, which was sufficient in his context (finite groups).

Permutations were studied by Joseph-Louis Lagrange in his 1770 paper Réflexions sur la résolution algébrique des équations (Thoughts on the algebraic solution of equations) devoted to solutions of algebraic equations, in which he introduced Lagrange resolvents. Lagrange's goal was to understand why equations of third and fourth degree admit formulas for solutions, and he identified as key objects permutations of the roots. An important novel step taken by Lagrange in this paper was the abstract view of the roots, i.e. as symbols and not as numbers. However, he did not consider composition of permutations. Serendipitously, the first edition of Edward Waring's Meditationes Algebraicae (Meditations on Algebra) appeared in the same year, with an expanded version published in 1782. Waring proved the fundamental theorem of symmetric polynomials, and specially considered the relation between the roots of a quartic equation and its resolvent cubic. Mémoire sur la résolution des équations (Memoire on the Solving of Equations) of Alexandre Vandermonde developed the theory of symmetric functions from a slightly different angle, but like Lagrange, with the goal of understanding solvability of algebraic equations.

Kronecker claimed in 1888 that the study of modern algebra began with this first paper of Vandermonde. Cauchy states quite clearly that Vandermonde had priority over Lagrange for this remarkable idea, which eventually led to the study of group theory.

Paolo Ruffini was the first person to develop the theory of permutation groups, and like his predecessors, also in the context of solving algebraic equations. His goal was to establish the impossibility of an algebraic solution to a general algebraic equation of degree greater than four. En route to this goal he introduced the notion of the order of an element of a group, conjugacy, the cycle decomposition of elements of permutation groups and the notions of primitive and imprimitive and proved some important theorems relating these concepts, such as,

If G is a subgroup of S_5 whose order is divisible by 5 then G contains an element of order 5.

However, that he got by without formalizing the concept of a group, or even of a permutation group. The next step was taken by Évariste Galois in 1832, although his work remained unpublished until

1846, when he considered for the first time what is now called the *closure property* of a group of permutations, which he expressed as:

If in such a group one has the substitutions S and T then one has the substitution ST.

The theory of permutation groups received further far-reaching development in the hands of Augustin Cauchy and Camille Jordan, both through introduction of new concepts and, primarily, a great wealth of results about special classes of permutation groups and even some general theorems. Among other things, Jordan defined a notion of isomorphism, still in the context of permutation groups and, incidentally, it was he who put the term *group* in wide use.

The abstract notion of a group appeared for the first time in Arthur Cayley's papers in 1854. Cayley realized that a group need not be a permutation group (or even *finite*), and may instead consist of matrices, whose algebraic properties, such as multiplication and inverses, he systematically investigated in succeeding years. Much later Cayley would revisit the question whether abstract groups were more general than permutation groups, and establish that, in fact, any group is isomorphic to a group of permutations.

Modern Algebra

The end of the 19th and the beginning of the 20th century saw a tremendous shift in the methodology of mathematics. Abstract algebra emerged around the start of the 20th century, under the name *modern algebra*. Its study was part of the drive for more intellectual rigor in mathematics. Initially, the assumptions in classical algebra, on which the whole of mathematics (and major parts of the natural sciences) depend, took the form of axiomatic systems. No longer satisfied with establishing properties of concrete objects, mathematicians started to turn their attention to general theory. Formal definitions of certain algebraic structures began to emerge in the 19th century. For example, results about various groups of permutations came to be seen as instances of general theorems that concern a general notion of an *abstract group*. Questions of structure and classification of various mathematical objects came to forefront.

These processes were occurring throughout all of mathematics, but became especially pronounced in algebra. Formal definition through primitive operations and axioms were proposed for many basic algebraic structures, such as groups, rings, and fields. Hence such things as group theory and ring theory took their places in pure mathematics. The algebraic investigations of general fields by Ernst Steinitz and of commutative and then general rings by David Hilbert, Emil Artin and Emmy Noether, building up on the work of Ernst Kummer, Leopold Kronecker and Richard Dedekind, who had considered ideals in commutative rings, and of Georg Frobenius and Issai Schur, concerning representation theory of groups, came to define abstract algebra. These developments of the last quarter of the 19th century and the first quarter of 20th century were systematically exposed in Bartel van der Waerden's *Moderne algebra*, the two-volume monograph published in 1930–1931 that forever changed for the mathematical world the meaning of the word *algebra* from *the theory of equations* to the *theory of algebraic structures*.

Basic Concepts

By abstracting away various amounts of detail, mathematicians have defined various algebraic structures that are used in many areas of mathematics. For instance, almost all systems studied are sets, to which the theorems of set theory apply. Those sets that have a certain binary operation defined on them form magmas, to which the concepts concerning magmas, as well those concerning sets, apply. We can add additional constraints on the algebraic structure, such as associativity (to form semigroups); identity, and inverses (to form groups); and other more complex structures. With additional structure, more theorems could be proved, but the generality is reduced. The "hierarchy" of algebraic objects (in terms of generality) creates a hierarchy of the corresponding theories: for instance, the theorems of group theory may be used when studying rings (algebraic objects that have two binary operations with certain axioms) since a ring is a group over one of its operations. In general there is a balance between the amount of generality and the richness of the theory: more general structures have usually fewer nontrivial theorems and fewer applications.

Examples of algebraic structures with a single binary operation are:

- Magma,
- Quasigroup,
- Monoid,
- Semigroup,
- Group.

Examples involving several operations include:

- Ring,
- Field,
- Module,
- Vector space,
- Algebra over a field,
- Associative algebra,
- Lie algebra,
- Lattice,
- Boolean algebra.

Applications

Because of its generality, abstract algebra is used in many fields of mathematics and science. For instance, algebraic topology uses algebraic objects to study topologies. The Poincaré conjecture, proved in 2003, asserts that the fundamental group of a manifold, which encodes information about connectedness, can be used to determine whether a manifold is a sphere or not. Algebraic number theory studies various number rings that generalize the set of integers. Using tools of algebraic number theory, Andrew Wiles proved Fermat's Last Theorem.

In physics, groups are used to represent symmetry operations, and the usage of group theory could simplify differential equations. In gauge theory, the requirement of local symmetry can be used to deduce the equations describing a system. The groups that describe those symmetries are Lie groups, and the study of Lie groups and Lie algebras reveals much about the physical system; for instance, the number of force carriers in a theory is equal to the dimension of the Lie algebra, and these bosons interact with the force they mediate if the Lie algebra is nonabelian.

Universal Algebra

Universal algebra studies common properties of all algebraic structures, including groups, rings, fields, lattices, etc. A universal algebra is a pair $A = (A, (f_i^A)_{(i \in I)})$, where A and I are sets and for each $i \in I$, f_i^A is an operation on A. The algebra A is finitary if each of its operations is finitary.

A set of function symbols (or operations) of degree $n \geq 0$ is called a signature (or type). Let Σ be a signature. An algebra A is defined by a domain S (which is called its carrier or universe) and a mapping that relates a function f: $S^n \to S$ to each n-place function symbol from Sigma.

Let A and B be two algebras over the same signature Sigma, and their carriers are A and B, respectively. A mapping $\phi : A \to B$ is called a homomorphism from A to B if for every $f \in \Sigma$ and all $x_1, \dots, x_n \in A$,

$$\phi(f(x_1, \dots, x_n)) = f(\phi(x_1), \dots, \phi(x_n)).$$

If a homomorphism ϕ is surjective, then it is called epimorphism. If ϕ is an epimorphism, then B is called a homomorphic image of A. If the homomorphism ϕ is a bijection, then it is called an isomorphism. On the class of all algebras, define a relation $\sim$ by $A \sim B$ if and only if there is an isomorphism from A onto B. Then the relation $\sim$ is an equivalence relation. Its equivalence classes are called isomorphism classes, and are typically proper classes.

A homomorphism from A to B is often denoted as $\phi : A \to B$. A homomorphism $\phi : A \to A$ is called an endomorphism. An isomorphism $\phi : A \to A$ is called an automorphism. The notions of homomorphism, isomorphism, endomorphism, etc., are generalizations of the respective notions in groups, rings, and other algebraic theories.

Identities (or equalities) in algebra A over signature Sigma have the form:

$$s = t,$$

where s and t are terms built up from variables using function symbols from Σ. An identity $s = t$ is said to hold in an algebra A if it is true for all possible values of variables in the identity, i.e., for all possible ways of replacing the variables by elements of the carrier. The algebra A is then said to satisfy the identity $s = t$.

References

- Algebra, science: britannica.com, Retrieved 18 March, 2019
- Elementary-algebra, science: britannica.com, Retrieved 19 April, 2019
- Schumm, bruce (2004), deep down things, baltimore: johns hopkins university press, isbn 0-8018-7971-x
- Universalalgebra: mathworld.wolfram.com, Retrieved 20 May, 2019

Trigonometry 5

- **Pythagorean Triple**
- **Pythagorean Theorem**
- **Trigonometric Functions**
- **Inverse Trigonometric Functions**
- **Applications of Trigonometry**

The branch of mathematics which deals with the relation of lines and angles in a triangle is referred to as trigonometry. Some of its basic principles are Pythagorean triple, Pythagorean theorem, trigonometric functions, inverse trigonometric functions, etc. All the diverse principles of trigonometry have been carefully analyzed in this chapter.

Trigonometry is the branch of mathematics concerned with specific functions of angles and their application to calculations. There are six functions of an angle commonly used in trigonometry. Their names and abbreviations are sine (sin), cosine (cos), tangent (tan), cotangent (cot), secant (sec), and cosecant (csc). These six trigonometric functions in relation to a right triangle are displayed in the figure. For example, the triangle contains an angle A, and the ratio of the side opposite to A and the side opposite to the right angle (the hypotenuse) is called the sine of A, or sin A; the other trigonometry functions are defined similarly. These functions are properties of the angle A independent of the size of the triangle, and calculated values were tabulated for many angles before computers made trigonometry tables obsolete. Trigonometric functions are used in obtaining unknown angles and distances from known or measured angles in geometric figures.

Trigonometry developed from a need to compute angles and distances in such fields as astronomy, mapmaking, surveying, and artillery range finding. Problems involving angles and distances in one plane are covered in plane trigonometry. Applications to similar problems in more than one plane of three-dimensional space are considered in spherical trigonometry.

Classical Trigonometry

Until about the 16th century, trigonometry was chiefly concerned with computing the numerical values of the missing parts of a triangle (or any shape that can be dissected into triangles) when the values of other parts were given. For example, if the lengths of two sides of a triangle and the measure of the enclosed angle are known, the third side and the two remaining angles can be calculated. Such calculations distinguish trigonometry from geometry, which mainly investigates qualitative relations. Of course, this distinction is not always absolute: the Pythagorean theorem, for example, is a statement about the lengths of the three sides in a right triangle and is thus quantitative in nature. Still, in its original form, trigonometry was by and large an offspring of geometry; it was not until the 16th century that the two became separate branches of mathematics.

Ancient Egypt and the Mediterranean World

Several ancient civilizations—in particular, the Egyptian, Babylonian, Hindu, and Chinese—possessed a considerable knowledge of practical geometry, including some concepts that were a prelude to trigonometry. The Rhind papyrus, an Egyptian collection of 84 problems in arithmetic, algebra, and geometry dating from about 1800 BCE, contains five problems dealing with the seked. A close analysis of the text, with its accompanying figures, reveals that this word means the slope of an incline—essential knowledge for huge construction projects such as the pyramids. For example, problem 56 asks: "If a pyramid is 250 cubits high and the side of its base is 360 cubits long, what is its seked?" The solution is given as 51/25 palms per cubit, and, since one cubit equals 7 palms, this fraction is equivalent to the pure ratio 18/25. This is actually the "run-to-rise" ratio of the pyramid in question—in effect, the cotangent of the angle between the base and face. It shows that the Egyptians had at least some knowledge of the numerical relations in a triangle, a kind of "proto-trigonometry."

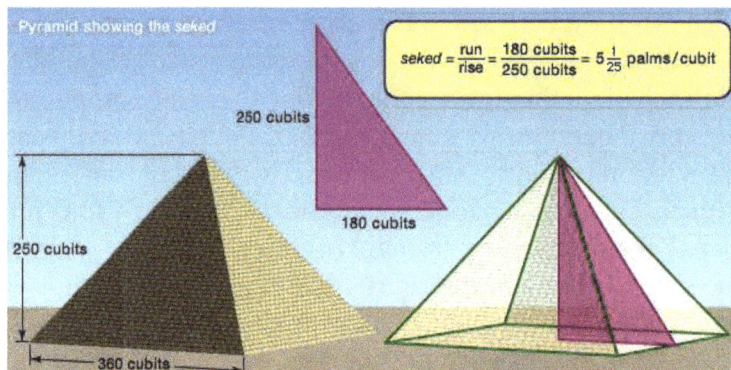

The Egyptians defined the seked as the ratio of the run to the rise,
which is the reciprocal of the modern definition of the slope.

Trigonometry in the modern sense began with the Greeks. Hipparchus was the first to construct a table of values for a trigonometric function. He considered every triangle—planar or spherical—as being inscribed in a circle, so that each side becomes a chord (that is, a straight line that connects two points on a curve or surface, as shown by the inscribed triangle ABC in the figure). To compute the various parts of the triangle, one has to find the length of each chord as a function of the central angle that subtends it—or, equivalently, the length of a chord as a function of the corresponding arc width. This became the chief task of trigonometry for the next several centuries. As an astronomer, Hipparchus was mainly interested in spherical triangles, such as the imaginary triangle formed by three stars on the celestial sphere, but he was also familiar with the basic formulas of plane trigonometry. In Hipparchus's time these formulas were expressed in purely geometric terms as relations between the various chords and the angles (or arcs) that subtend them; the modern symbols for the trigonometric functions were not introduced until the 17th century.

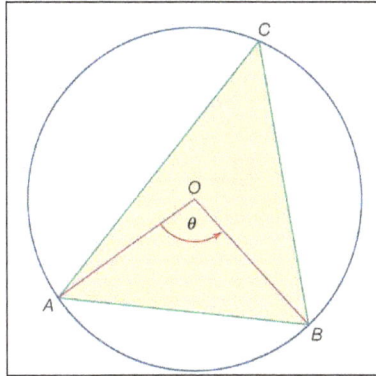

Triangle inscribed in a circle this figure illustrates the relationship between a central angle θ (an angle formed by two radii in a circle) and its chord AB (equal to one side of an inscribed triangle).

Common Trigonometry Formulas

Variations on the Pythagorean theorem:

$$\sin^2 A + \cos^2 A = 1$$
$$\tan^2 A + 1 = \sec^2 A$$
$$1 + \cot^2 A = \cos^2 A$$

Half − angle formulas:

$$\sin^2\left(\frac{A}{2}\right) = \frac{1 - \cos A}{2}$$
$$\cos^2\left(\frac{A}{2}\right) = \frac{1 + \cos A}{2}$$

Double − angle formules:

$$\sin(2A) = 2\sin A \cos A$$
$$\cos(2A) = \cos^2 A - \sin^2 A$$

Additoion formules:

$$\sin(A \pm B) = \sin A \cos B \pm \cos A \sin B$$
$$\cos(A \pm B) = \cos A \cos B \mp \sin A \sin B$$

Law of sin es:

$$\frac{a}{\sin A} = \frac{b}{\sin B} = \frac{c}{\sin C}$$

Law of cos ines:

$$c^2 = a^2 + b^2 - 2ab \cos C$$
$$b^2 = a^2 + c^2 - 2ac \cos B$$
$$a^2 = b^2 + c^2 - 2bc \cos A$$

The first major ancient work on trigonometry to reach Europe intact after the Dark Ages was the Almagest by Ptolemy. He lived in Alexandria, the intellectual centre of the Hellenistic world, but little else is known about him. Although Ptolemy wrote works on mathematics, geography, and optics, he is chiefly known for the Almagest, a 13-book compendium on astronomy that became the basis for humankind's world picture until the heliocentric system of Nicolaus Copernicus began to supplant Ptolemy's geocentric system in the mid-16th century. In order to develop this world picture—the essence of which was a stationary Earth around which the Sun, Moon, and the five known planets move in circular orbits—Ptolemy had to use some elementary trigonometry. This is essentially a table of sines, which can be seen by denoting the radius r, the arc A, and the length of the subtended chord c, to obtain c = 2r sin A/2. Because Ptolemy used the Babylonian sexagesimal numerals and numeral systems (base 60), he did his computations with a standard circle of radius r = 60 units, so that c = 120 sin A/2. Thus, apart from the proportionality factor 120, his was a table of values of sin A/2 and therefore (by doubling the arc) of sin A. With the help of his table Ptolemy improved on existing geodetic measures of the world and refined Hipparchus's model of the motions of the heavenly bodies.

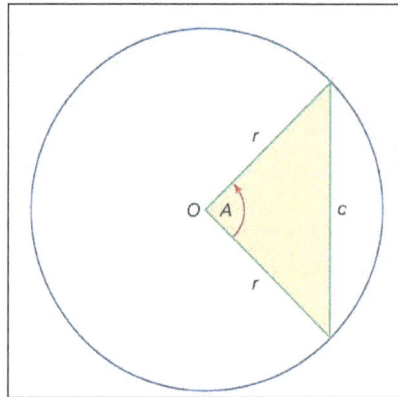

Constructing a table of chords: By labeling the central angle A, the radii r, and the chord c, it can be shown that c = 2r sin (A/2).

India and the Islamic World

The next major contribution to trigonometry came from India. In the sexagesimal system, multiplication or division by 120 (twice 60) is analogous to multiplication or division by 20 (twice 10) in the decimal system. Thus, rewriting Ptolemy's formula as $^c/_{120} = sin B$, where $B = {}^A/_2$, the relation expresses the half-chord as a function of the arc B that subtends it—precisely the modern sine function. The first table of sines is found in the Aryabhatiya. Its author, Aryabhata I, used the word ardha-jya for half-chord, which he sometimes turned around to jya-ardha ("chord-half"); in due time he shortened it to jya or jiva.

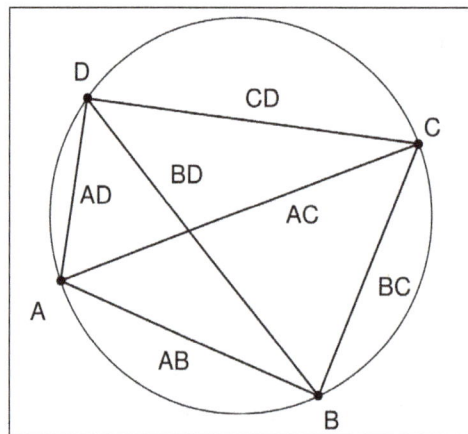

Ptolemy's formula.

During the Middle Ages, while Europe was plunged into darkness, the torch of learning was kept alive by Arab and Jewish scholars living in Spain, Mesopotamia, and Persia. The first table of tangents and cotangents was constructed around 860 by Ḥabash al-Ḥāsib ("the Calculator"), who wrote on astronomy and astronomical instruments. Another Arab astronomer, al-Battāni, gave a rule for finding the elevation θ of the Sun above the horizon in terms of the length s of the shadow cast by a vertical gnomon of height h. Al-Battāni's rule, s = h sin (90° − θ)/sin θ, is equivalent to the formula s = h cot θ. Based on this rule he constructed a "table of shadows"—essentially a table of cotangents—for each degree from 1° to 90°. It was through al-Battāni's work that the Hindu half-chord function—equivalent to the modern sine—became known in Europe.

Passage to Europe

Until the 16th century it was chiefly spherical trigonometry that interested scholars—a consequence of the predominance of astronomy among the natural sciences. The first definition of a spherical triangle is contained in Book 1 of the Sphaerica, a three-book treatise by Menelaus of Alexandria in which Menelaus developed the spherical equivalents of Euclid's propositions for planar triangles. A spherical triangle was understood to mean a figure formed on the surface of a sphere by three arcs of great circles, that is, circles whose centres coincide with the centre of the sphere. There are several fundamental differences between planar and spherical triangles. For example, two spherical triangles whose angles are equal in pairs are congruent (identical in size as well as in shape), whereas they are only similar (identical in shape) for the planar case. Also, the sum of the angles of a spherical triangle is always greater than 180°, in contrast to the planar case where the angles always sum to exactly 180°.

Several Arab scholars, notably Naṣīr al-Dīn al-Ṭūsī and al-Bāttāni, continued to develop spherical trigonometry and brought it to its present form. Ṭūsī was the first to write a work on trigonometry independently of astronomy. But the first modern book devoted entirely to trigonometry appeared in the Bavarian city of Nürnberg in 1533 under the title On Triangles of Every Kind. Its author was the astronomer Regiomontanus. On Triangles contains all the theorems needed to solve triangles, planar or spherical—although these theorems are expressed in verbal form, as symbolic algebra had yet to be invented. In particular, the law of sines is stated in essentially the modern way. On Triangles was greatly admired by future generations of scientists; the astronomer Nicolaus Copernicus studied it thoroughly, and his annotated copy survives.

The final major development in classical trigonometry was the invention of logarithms by the Scottish mathematician John Napier in 1614. His tables of logarithms greatly facilitated the art of numerical computation—including the compilation of trigonometry tables—and were hailed as one of the greatest contributions to science.

Modern Trigonometry

From Geometric to Analytic Trigonometry

In the 16th century trigonometry began to change its character from a purely geometric discipline to an algebraic-analytic subject. Two developments spurred this transformation: the rise of symbolic algebra, pioneered by the French mathematician François Viète, and the invention of analytic geometry by two other Frenchmen, Pierre de Fermat and René Descartes. Viète showed that the solution of many algebraic equations could be expressed by the use of trigonometric expressions. For example, the equation $x^3 = 1$ has the three solutions:

$$x = 1,$$

$$\cos 120° + i \sin 120° = \frac{-1 - i\sqrt{3}}{2}, \; and$$

$$\cos 240° + i \sin 240° = \frac{-1 - i\sqrt{3}}{2}$$

(Here i is the symbol for Square root of $\sqrt{-1}$, the "imaginary unit.") That trigonometric expressions may appear in the solution of a purely algebraic equation was a novelty in Viète's time; he used it to advantage in a famous encounter between King Henry IV of France and the Netherlands' ambassador to France. The latter spoke disdainfully of the poor quality of French mathematicians and challenged the king with a problem posed by Adriaen van Roomen, professor of mathematics and medicine at the University of Leuven (Belgium), to solve a certain algebraic equation of degree 45. The king summoned Viète, who immediately found one solution and on the following day came up with 22 more.

Viète was also the first to legitimize the use of infinite processes in mathematics. In 1593 he discovered the infinite product,

$$\frac{2}{\Pi} = \frac{\sqrt{2}}{2} \cdot \frac{\sqrt{(2+\sqrt{2}}}{2} \cdot \frac{\sqrt{(2+\sqrt{(2+\sqrt{2})})}}{2}$$

which is regarded as one of the most beautiful formulas in mathematics for its recursive pattern. By computing more and more terms, one can use this formula to approximate the value of π to any desired accuracy. In 1671 James Gregory found the power series.

Power series for three trigonometry functions:

$$\sin x = x - \frac{x^3}{3!} + \frac{x^5}{5!} - \frac{x^7}{7!} + \cdots$$

$$\cos x = 1 - \frac{x^2}{2!} + \frac{x^4}{4!} - \frac{x^6}{7!} + \cdots$$

$$\tan^{-1} x = x - \frac{x^3}{3} + \frac{x^5}{5} - \frac{x^7}{7} + \cdots$$

For the inverse tangent function (arc tan, or tan−1), from which he got, by letting x = 1, the formula:

$$\pi/_4 = 1 - ^1/_3 + ^1/_5 - ^1/_7 + \cdots,$$

which demonstrated a remarkable connection between π and the integers. Although the series converged too slowly for a practical computation of π (it would require 628 terms to obtain just two accurate decimal places). This was soon followed by Isaac Newton's discovery of the power series for sine and cosine. Research, however, has brought to light that some of these formulas were already known, in verbal form, by the Indian astronomer Madhava.

The gradual unification of trigonometry and algebra—and in particular the use of complex numbers (numbers of the form x + iy, where x and y are real numbers and i = $\sqrt{-1}$) in trigonometric expressions—was completed in the 18th century. In 1722 Abraham de Moivre derived, in implicit form, the famous formula:

$$(cos\, ø + i\, sin\, ø)n = cos\, nø + i\, sin\, nø,$$

which allows one to find the nth root of any complex number. It was the Swiss mathematician Leonhard Euler, though, who fully incorporated complex numbers into trigonometry. Euler's formula

$e^{i\phi} = cos\ ø + i\,sin\ ø$, where $e \cong 2.71828$ is the base of natural logarithms, appeared in 1748 in his great work Introductio in analysin infinitorum—although Roger Cotes already knew the formula in its inverse form $øi = log\ (cos\ ø + i\,sin\ ø)$ in 1714. Substituting into this formula the value $ø = \Pi$, one obtains ei $e^{i\pi} = cos\ \pi + i\,sin\ \pi = -1 + 0i = -1$ or equivalently, $e^{i\pi} + 1 = 0$. This most intriguing of all mathematical formulas contains the additive and multiplicative identities (0 and 1, respectively), the two irrational numbers that occur most frequently in the physical world (π and e), and the imaginary unit (i), and it also employs the basic operations of addition and exponentiation—hence its great aesthetic appeal. Finally, by combining his formula with its companion formula:

$$e^{-i\phi} = cos(-ø) + i\,sin\ (-ø) = cos\ ø - i\,sin\ ø,$$

Euler obtained the expressions:

$$cos\ ø = \frac{e^{i\phi} + e^{-\phi}}{2} \qquad sin\ ø = \frac{e^{i\phi} - e^{-i\phi}}{2i}$$

which are the basis of modern analytic trigonometry.

Application to Science

While these developments shifted trigonometry away from its original connection to triangles, the practical aspects of the subject were not neglected. The 17th and 18th centuries saw the invention of numerous mechanical devices—from accurate clocks and navigational tools to musical instruments of superior quality and greater tonal range—all of which required at least some knowledge of trigonometry. A notable application was the science of artillery—and in the 18th century it was a science. Galileo Galilei discovered that any motion—such as that of a projectile under the force of gravity—can be resolved into two components, one horizontal and the other vertical, and that these components can be treated independently of one another. This discovery led scientists to the formula for the range of a cannonball when its muzzle velocity vo (the speed at which it leaves the cannon) and the angle of elevation A of the cannon are given. The theoretical range, in the absence of air resistance, is given by:

$$R = \frac{v_0^{\,2}\ sin\,2A}{g}$$

where g is the acceleration due to gravity (about 9.81 metres/second²). This formula shows that, for a given muzzle velocity, the range depends solely on A; it reaches its maximum value when A = 45° and falls off symmetrically on either side of 45°. These facts, of course, had been known empirically for many years, but their theoretical explanation was a novelty in Galileo's time.

Another practical aspect of trigonometry that received a great deal of attention during this time period was surveying. The method of triangulation was first suggested in 1533 by the Dutch mathematician Gemma Frisius: one chooses a base line of known length, and from its endpoints the angles of sight to a remote object are measured. The distance to the object from either endpoint can then be calculated by using elementary trigonometry. The process is then repeated with the new distances as base lines, until the entire area to be surveyed is covered by a network of triangles. The method was first carried out on a large scale by another Dutchman, Willebrord Snell, who surveyed a stretch of 130 km (80 miles) in Holland, using 33 triangles. The French government, under the leadership of the astronomer Jean Picard, undertook to triangulate the entire country, a task that was to take over a century and involve four generations of the Cassini family (Gian, Jacques, César-François, and Dominique) of astronomers. The British undertook an even more ambitious task—the survey of the entire subcontinent of India. Known as the Great Trigonometric Survey, it lasted from 1800 to 1913 and culminated with the discovery of the tallest mountain on Earth—Peak XV, or Mount Everest.

Concurrent with these developments, 18th-century scientists also turned their attention to aspects of the trigonometric functions that arose from their periodicity. If the cosine and sine functions are defined as the projections on the x- and y-axes, respectively, of a point moving on a unit circle (a circle with its centre at the origin and a radius of 1), then these functions will repeat their values every 360°, or 2π radians. Hence the importance of the sine and cosine functions in describing periodic phenomena—the vibrations of a violin string, the oscillations of a clock pendulum, or the propagation of electromagnetic waves. These investigations reached a climax when Joseph Fourier discovered that almost any periodic function can be expressed as an infinite sum of sine and cosine functions, whose periods are integral divisors of the period of the original function. For example, the "sawtooth" function can be written as:

$$2(\sin x - \frac{\sin 2x}{2} + \frac{\sin 3x}{3} - \cdots);$$

as successive terms in the series are added, an ever-better approximation to the sawtooth function results. These trigonometric or Fourier series have found numerous applications in almost every branch of science, from optics and acoustics to radio transmission and earthquake analysis. Their extension to nonperiodic functions played a key role in the development of quantum mechanics in the early years of the 20th century. Trigonometry, by and large, matured with Fourier's theorem; further developments.

Principles of Trigonometry

Trigonometric Functions

A somewhat more general concept of angle is required for trigonometry than for geometry. An angle A with vertex at V, the initial side of which is VP and the terminal side of which is VQ, is indicated in the figure by the solid circular arc. This angle is generated by the continuous counterclockwise rotation of a line segment about the point V from the position VP to the position VQ. A second angle A′ with the same initial and terminal sides, indicated in the figure by the broken circular arc, is generated by the clockwise rotation of the line segment from the position VP to the position VQ. Angles are considered positive when generated by counterclockwise rotations, negative when generated by clockwise rotations. The positive angle A and the negative angle A′ in the figure are generated by less than one complete rotation of the line segment about the point V. All other positive and negative angles with the same initial and terminal sides are obtained by rotating the line segment one or more complete turns before coming to rest at VQ.

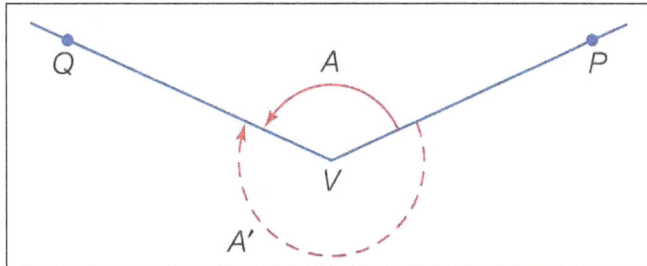

This figure shows a positive general angle A, as well as a negative general angle A'.

Numerical values can be assigned to angles by selecting a unit of measure. The most common units are the degree and the radian. There are 360° in a complete revolution, with each degree further divided into 60′ (minutes) and each minute divided into 60″ (seconds). In theoretical work, the radian is the most convenient unit. It is the angle at the centre of a circle that intercepts an arc equal in length to the radius; simply put, there are 2Π radians in one complete revolution. From these definitions, it follows that $1° = {}^{\pi}/_{180}$ radians.

Equal angles are angles with the same measure; i.e., they have the same sign and the same number of degrees. Any angle −A has the same number of degrees as A but is of opposite sign. Its measure, therefore, is the negative of the measure of A. If two angles, A and B, have the initial sides VP and VQ and the terminal sides VQ and VR, respectively, then the angle A + B has the initial and terminal sides VP and VR. The angle A + B is called the sum of the angles A and B, and its relation to A and B when A is positive and B is positive or negative is illustrated in the figure. The sum A + B is the angle the measure of which is the algebraic sum of the measures of A and B. The difference A − B is the sum of A and −B. Thus, all angles coterminal with angle A (i.e., with the same initial and terminal sides as angle A) are given by A ± 360n, in which 360n is an angle of n complete revolutions. The angles (180 − A) and (90 − A) are the supplement and complement of angle A, respectively.

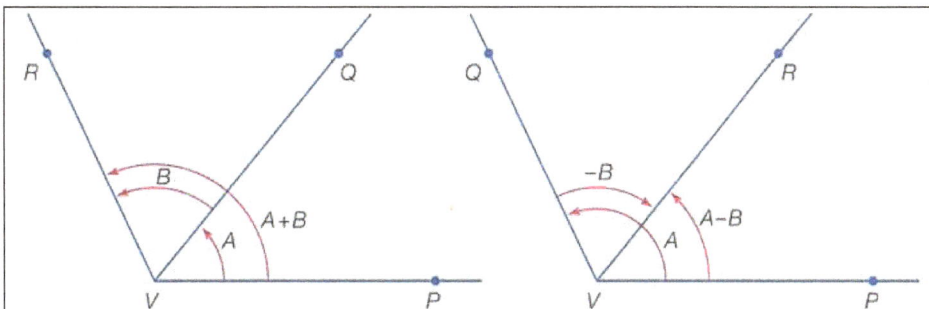

Addition of angles: The figure indicates how to add a positive or negative angle (B) to a positive angle (A).

Trigonometric Functions of an Angle

To define trigonometric functions for any angle A, the angle is placed in position on a rectangular coordinate system with the vertex of A at the origin and the initial side of A along the positive x-axis; r (positive) is the distance from V to any point Q on the terminal side of A, and (x, y) are the rectangular coordinates of Q.

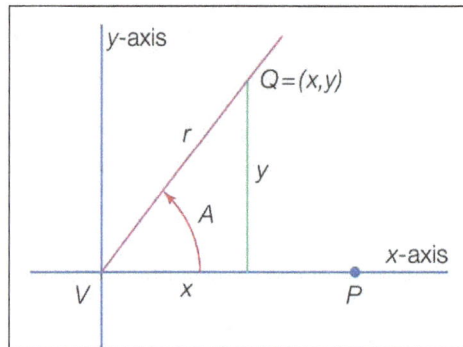

Angle in standard position: The figure shows an angle A in standard position, that is, with initial side on the x-axis.

The six functions of A are then defined by six ratios exactly as in the earlier case for the triangle given in the introduction. Because division by zero is not allowed, the tangent and secant are not defined for angles the terminal side of which falls on the y-axis, and the cotangent and cosecant are undefined for angles the terminal side of which falls on the x-axis. When the Pythagorean equality $x^2 + y^2 = r^2$ is divided in turn by r^2, x^2, and y^2, the three squared relations relating cosine and sine, tangent and secant, cotangent and cosecant are obtained.

Negative angles:

$$\sin(-A) = -\sin A \qquad \csc(-A) = -\csc A$$
$$\cos(-A) = \cos A \qquad \sec(-A) = \sec A$$
$$\tan(-A) = -\tan A \qquad \cot(-A) = -\cot A$$

If the point Q on the terminal side of angle A in standard position has coordinates (x, y), this point will have coordinates (x, −y) when on the terminal side of −A in standard position. From this fact and the definitions are obtained further identities for negative angles. These relations may also be stated briefly by saying that cosine and secant are even functions (symmetrical about the y-axis), while the other four are odd functions (symmetrical about the origin).

It is evident that a trigonometric function has the same value for all coterminal angles. When n is an integer, therefore, sin (A ± 360n) = sin A; there are similar relations for the other five functions. These results may be expressed by saying that the trigonometric functions are periodic and have a period of 360° or 180°.

Complementary angles and cofunctions:

$$\sin(A \pm 90°) = \pm\cos A \qquad \csc(A \pm 90°) = \pm\sec A$$
$$\cos(A \pm 90°) = \mp\sin A \qquad \sec(A \pm 90°) = \mp\csc A$$
$$\tan(A \pm 90°) = -\cot A \qquad \cot(A \pm 90°) = -\tan A$$

When Q on the terminal side of A in standard position has coordinates (x, y), it has coordinates (−y, x) and (y, −x) on the terminal side of A + 90 and A − 90 in standard position, respectively. Consequently, six formulas equate a function of the complement of A to the corresponding cofunction of A.

Tables of Natural Functions

To be of practical use, the values of the trigonometric functions must be readily available for any given angle. Various trigonometric identities show that the values of the functions for all angles can readily be found from the values for angles from 0° to 45°. For this reason, it is sufficient to list in a table the values of sine, cosine, and tangent for all angles from 0° to 45° that are integral multiples of some convenient unit (commonly 1′). Before computers rendered them obsolete in the late 20th century, such trigonometry tables were helpful to astronomers, surveyors, and engineers.

Table: Common angles for trigonometry functions.

	0°	30°	45°	60°	90°
sin	0	$1/2$	$\sqrt{2}/2$	$\sqrt{3}/2$	1
cos	1	$\sqrt{3}/2$	$\sqrt{2}/2$	$1/2$	0
tan	0	$\sqrt{3}/3$	1	$\sqrt{3}$	*underfined*

For angles that are not integral multiples of the unit, the values of the functions may be interpolated. Because the values of the functions are in general irrational numbers, they are entered in the table as decimals, rounded off at some convenient place. For most purposes, four or five decimal places are sufficient, and tables of this accuracy are common. Simple geometrical facts alone, however, suffice to determine the values of the trigonometric functions for the angles 0°, 30°, 45°, 60°, and 90°. These values are listed in a table for the sine, cosine, and tangent functions.

Plane Trigonometry

In many applications of trigonometry the essential problem is the solution of triangles. If enough sides and angles are known, the remaining sides and angles as well as the area can be calculated, and the triangle is then said to be solved. Triangles can be solved by the law of sines and the law of cosines. To secure symmetry in the writing of these laws, the angles of the triangle are lettered A, B, and C and the lengths of the sides opposite the angles are lettered a, b, and c, respectively.

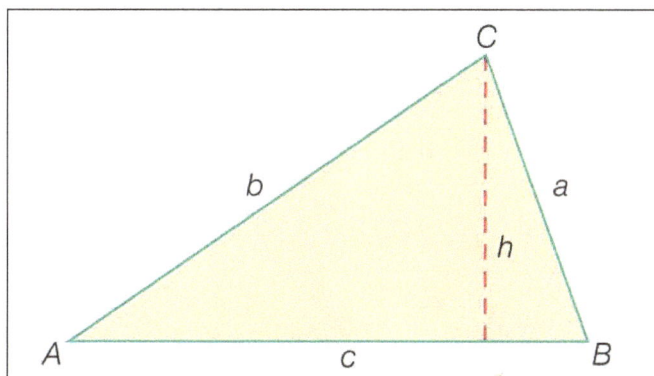

Standard lettering of a triangle: In addition to the angles (A, B, C) and sides (a, b, c), one of the three heights of the triangle (h) is included by drawing the line segment from one of the triangle's vertices (in this case C) that is perpendicular to the opposite side of the triangle.

The law of sines is expressed as an equality involving three sine functions while the law of cosines is an identification of the cosine with an algebraic expression formed from the lengths of sides opposite the corresponding angles. To solve a triangle, all the known values are substituted into equations expressing the laws of sines and cosines, and the equations are solved for the unknown quantities. For example, the law of sines is employed when two angles and a side are known or when two sides and an angle opposite one are known. Similarly, the law of cosines is appropriate when two sides and an included angle are known or three sides are known. Texts on trigonometry derive other formulas for solving triangles and for checking the solution. Older textbooks frequently included formulas especially suited to logarithmic calculation. Newer textbooks, however, frequently include simple computer instructions for use with a symbolic mathematical program.

Spherical Trigonometry

Spherical trigonometry involves the study of spherical triangles, which are formed by the intersection of three great circle arcs on the surface of a sphere. Spherical triangles were subject to intense study from antiquity because of their usefulness in navigation, cartography, and astronomy. The angles of a spherical triangle are defined by the angle of intersection of the corresponding tangent lines to each vertex. The sum of the angles of a spherical triangle is always greater than the sum of the angles in a planar triangle (π radians, equivalent to two right angles). The amount by which each spherical triangle exceeds two right angles (in radians) is known as its spherical excess. The area of a spherical triangle is given by the product of its spherical excess E and the square of the radius r of the sphere it resides on—in symbols, Er^2.

Common spherical trigonometry formulas:

$$\text{Law of sines:} \quad \frac{\sin a}{\sin A} = \frac{\sin b}{\sin B} = \frac{\sin c}{\sin C}$$

$$\text{Law of cosines:} \quad \begin{aligned} \cos a &= \cos b \cos c + \sin b \sin c \cos A \\ \cos b &= \cos a \cos c + \sin a \sin c \cos B \\ \cos c &= \cos a \cos b + \sin a \sin b \cos C \end{aligned}$$

$$\text{Half-angle formulas:} \quad \begin{aligned} \tan\left(\frac{A}{2}\right) &= \sqrt{\frac{\sin(s-b)\sin(s-c)}{\sin s \sin(s-a)}} \\ \tan\left(\frac{B}{2}\right) &= \sqrt{\frac{\sin(s-c)\sin(s-a)}{\sin s \sin(s-b)}} \\ \tan\left(\frac{C}{2}\right) &= \sqrt{\frac{\sin(s-a)\sin(s-b)}{\sin s \sin(s-c)}}, \text{ where } s = \frac{a+b+c}{2} \end{aligned}$$

$$\text{Half-side formulas:} \quad \begin{aligned} \tan\left(\frac{a}{2}\right) &= \sqrt{\frac{-\cos S \cos(S-A)}{\cos(S-B)\cos(S-C)}} \\ \tan\left(\frac{b}{2}\right) &= \sqrt{\frac{-\cos S \cos(S-B)}{\cos(S-A)\cos(S-C)}} \\ \tan\left(\frac{c}{2}\right) &= \sqrt{\frac{-\cos S \cos(S-C)}{\cos(S-A)\cos(S-B)}}, \text{ where } S = \frac{A+B+C}{2} \end{aligned}$$

By connecting the vertices of a spherical triangle with the centre O of the sphere that it resides on, a special "angle" known as a trihedral angle is formed. The central angles (also known as dihedral angles) between each pair of line segments OA, OB, and OC are labeled α, β, and γ to correspond to the sides (arcs) of the spherical triangle labeled a, b, and c, respectively. Because a trigonometric function of a central angle and its corresponding arc have the same value, spherical trigonometry formulas are given in terms of the spherical angles A, B, and C and, interchangeably, in terms of the arcs a, b, and c and the dihedral angles α, β, and γ. Furthermore, most formulas from plane trigonometry have an analogous representation in spherical trigonometry. For example, there is a spherical law of sines and a spherical law of cosines.

As was described for a plane triangle, the known values involving a spherical triangle are substituted in the analogous spherical trigonometry formulas, such as the laws of sines and cosines, and the resulting equations are then solved for the unknown quantities.

Napier's analogies:

$$\tan\left(\frac{a}{2}\right)\cos\left(\frac{B-C}{2}\right) = \tan\left(\frac{b+c}{2}\right)\cos\left(\frac{B+C}{2}\right)$$

$$\tan\left(\frac{a}{2}\right)\sin\left(\frac{B-C}{2}\right) = \tan\left(\frac{b-c}{2}\right)\sin\left(\frac{B+C}{2}\right)$$

$$\cot\left(\frac{A}{2}\right)\cos\left(\frac{b-c}{2}\right) = \tan\left(\frac{B+C}{2}\right)\cos\left(\frac{b+c}{2}\right)$$

$$\cot\left(\frac{A}{2}\right)\sin\left(\frac{b-c}{2}\right) = \tan\left(\frac{B-C}{2}\right)\sin\left(\frac{b+c}{2}\right)$$

Many other relations exist between the sides and angles of a spherical triangle. Worth mentioning are Napier's analogies (derivable from the spherical trigonometry half-angle or half-side formulas), which are particularly well suited for use with logarithmic tables.

Analytic Trigonometry

Analytic trigonometry combines the use of a coordinate system, such as the Cartesian coordinate system used in analytic geometry, with algebraic manipulation of the various trigonometry functions to obtain formulas useful for scientific and engineering applications.

Trigonometric functions of a real variable x are defined by means of the trigonometric functions of an angle. For example, sin x in which x is a real number is defined to have the value of the sine of the angle containing x radians. Similar definitions are made for the other five trigonometric functions of the real variable x. These functions satisfy the previously noted trigonometric relations with A, B, 90°, and 360° replaced by $x, y, {}^{\Pi}/_{2}$ radians, and 2Π radians, respectively. The minimum period of tan x and cot x is Π, and of the other four functions it is 2Π.

In calculus it is shown that sin x and cos x are sums of power series. These series may be used to compute the sine and cosine of any angle. For example, to compute the sine of 10°, it is necessary to find the value of $sin\,{}^{\Pi}/_{18}$ because 10° is the angle containing ${}^{\Pi}/_{18}$ radians. When ${}^{\Pi}/_{18}$ is substituted in the series for sin x, it is found that the first two terms give 0.17365, which is correct to five decimal places for the sine of 10°. By taking enough terms of the series, any number of decimal places can be correctly obtained. Tables of the functions may be used to sketch the graphs of the functions.

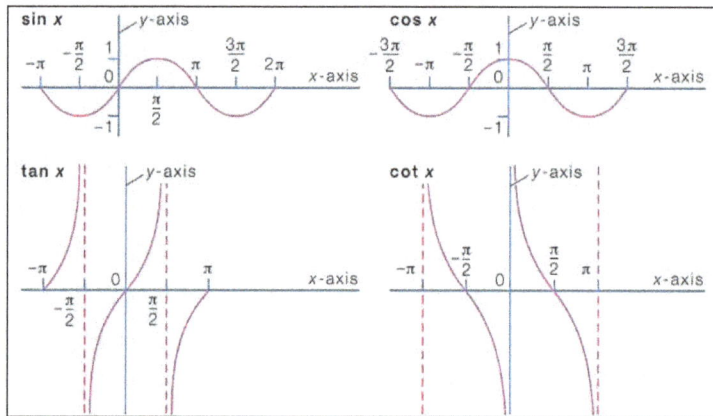

Graphs of some trigonometric functions. Note that each of these functions is periodic. Thus, the sine and cosine functions repeat every 2π, and the tangent and cotangent functions repeat every π.

Each trigonometric function has an inverse function, that is, a function that "undoes" the original function. For example, the inverse function for the sine function is written arcsin or sin^{-1}, thus $sin^{-1}(sin x) = sin(sin^{-1} x) = x$. The other trigonometric inverse functions are defined similarly.

Coordinates and Transformation of Coordinates

Polar Coordinates

For problems involving directions from a fixed origin (or pole) O, it is often convenient to specify a point P by its polar coordinates (r, θ), in which r is the distance OP and θ is the angle that the direction of r makes with a given initial line. The initial line may be identified with the x-axis of rectangular Cartesian coordinates, as shown in the figure. The point (r, θ) is the same as $(r, \theta + 2n\pi)$ for any integer n. It is sometimes desirable to allow r to be negative, so that (r, θ) is the same as $(-r, \theta + \pi)$.

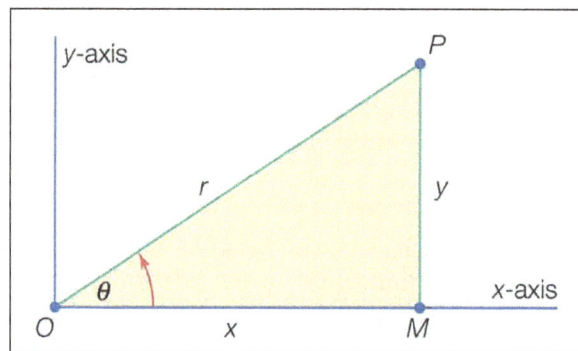

Cartesian and polar coordinates: The point labeled P in the figure resides in the plane. Therefore, it requires two dimensions to fix its location, either in Cartesian coordinates (x, y) or in polar coordinates (r, θ).

Given the Cartesian equation for a curve, the polar equation for the same curve can be obtained in terms of the radius r and the angle θ by substituting r cos θ and r sin θ for x and y, respectively. For example, the circle $x^2 + y^2 = a^2$ has the polar equation $(r\cos \theta)^2 + (r\sin \theta)^2 = a^2$, which reduces to r = a. (The positive value of r is sufficient, if θ takes all values from $-\pi$ to π or from o to 2π). Thus the polar equation of a circle simply expresses the fact that the curve is independent of θ and has constant radius. In a similar manner, the line $y = x \tan \phi$ has the polar equation

$\sin \theta = \cos \theta \tan \phi$, which reduces to $\theta = \phi$. (The other solution, $\theta = \phi + \Pi$, can be discarded if r is allowed to take negative values).

Transformation of Coordinates

A transformation of coordinates in a plane is a change from one coordinate system to another. Thus, a point in the plane will have two sets of coordinates giving its position with respect to the two coordinate systems used, and a transformation will express the relationship between the coordinate systems. For example, the transformation between polar and Cartesian coordinates discussed in the preceding section is given by x = r cos θ and y = r sin θ. Similarly, it is possible to accomplish transformations between rectangular and oblique coordinates.

In a translation of Cartesian coordinate axes, a transformation is made between two sets of axes that are parallel to each other but have their origins at different positions. If a point P has coordinates (x, y) in one system, its coordinates in the second system are given by (x – h, y – k) where (h, k) is the origin of the second system in terms of the first coordinate system. Thus, the transformation of P between the first system (x, y) and the second system (x′, y′) is given by the equations x = x′ + h and y = y′ + k. The common use of translations of axes is to simplify the equations of curves. For example, the equation $2x^2 + y^2 - 12x - 2y + 17 = 0$ can be simplified with the translations x′ = x – 3 and y′ = y – 1 to an equation involving only squares of the variables and a constant term: $(x')^2 + (y')^2/2 = 1$. In other words, the curve represents an ellipse with its centre at the point (3, 1) in the original coordinate system.

A rotation of coordinate axes is one in which a pair of axes giving the coordinates of a point (x, y) rotate through an angle ϕ to give a new pair of axes in which the point has coordinates (x′, y′), as shown in the figure. The transformation equations for such a rotation are given by $x = x' \cos \phi - y' \sin \phi$ and $y = x' \sin \phi + y' \cos \phi$. The application of these formulas with $\phi = 45°$ to the difference of squares, $x^2 - y^2 = a^2$, leads to the equation x′y′ = c (where c is a constant that depends on the value of a). This equation gives the form of the rectangular hyperbola when its asymptotes (the lines that a curve approaches without ever quite meeting) are used as the coordinate axes.

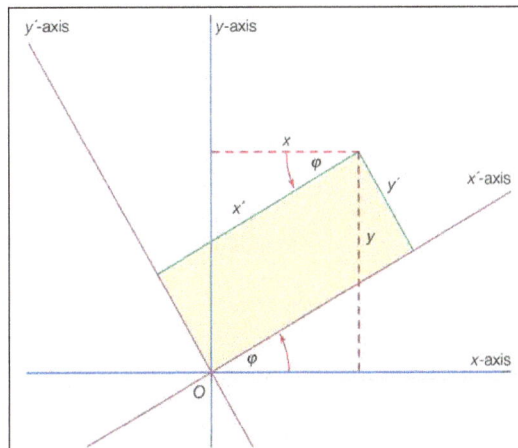

Rotation of axes: Rotating the coordinate axes through an angle ϕ changes the coordinates of a point from (x, y) to (x′, y′).

Pythagorean Triple

A Pythagorean triple consists of three positive integers a, b, and c, such that $a^2 + b^2 = c^2$. Such a triple is commonly written (a, b, c), and a well-known example is $(3, 4, 5)$. If (a, b, c) is a Pythagorean triple, then so is (ka, kb, kc) for any positive integer k. A primitive Pythagorean triple is one in which a, b and c are coprime (that is, they have no common divisor larger than 1). A triangle whose sides form a Pythagorean triple is called a Pythagorean triangle, and is necessarily a right triangle.

The name is derived from the Pythagorean theorem, stating that every right triangle has side lengths satisfying the formula $a^2 + b^2 = c^2$; thus, Pythagorean triples describe the three integer side lengths of a right triangle. However, right triangles with non-integer sides do not form Pythagorean triples. For instance, the triangle with sides $a = b = 1$ and $c = \sqrt{2}$ is a right triangle, but $(1, 1, \sqrt{2})$ is not a Pythagorean triple because $\sqrt{2}$ is not an integer. Moreover, 1 and $\sqrt{2}$ do not have an integer common multiple because $\sqrt{2}$ is irrational.

Pythagorean triples have been known since ancient times. The oldest known record comes from Plimpton 322, a Babylonian clay tablet from about 1800 BC, written in a sexagesimal number system. It was discovered by Edgar James Banks shortly after 1900, and sold to George Arthur Plimpton in 1922, for $10.

When searching for integer solutions, the equation $a^2 + b^2 = c^2$ is a Diophantine equation. Thus Pythagorean triples are among the oldest known solutions of a nonlinear Diophantine equation.

Examples:

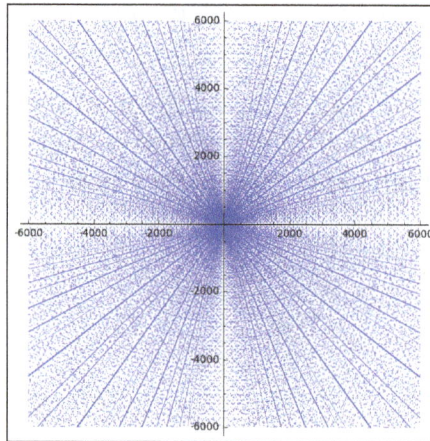

Scatter plot of the legs (a,b) of the first Pythagorean triples with a and b less than 6000.
Negative values are included to illustrate the parabolic patterns.

Table: 16 primitive Pythagorean triples with $c \leq 100$.

(3, 4, 5)	(5, 12, 13)	(8, 15, 17)	(7, 24, 25)
(20, 21, 29)	(12, 35, 37)	(9, 40, 41)	(28, 45, 53)
(11, 60, 61)	(16, 63, 65)	(33, 56, 65)	(48, 55, 73)
(13, 84, 85)	(36, 77, 85)	(39, 80, 89)	(65, 72, 97)

For example, that $(6, 8, 10)$ is *not* a primitive Pythagorean triple, as it is a multiple of $(3, 4, 5)$. Each of these low-c points forms one of the more easily recognizable radiating lines in the scatter plot.

Table: Additionally these are all the primitive Pythagorean triples with $100 < c \leq 300$.

(20, 99, 101)	(60, 91, 109)	(15, 112, 113)	(44, 117, 125)
(88, 105, 137)	(17, 144, 145)	(24, 143, 145)	(51, 140, 149)
(85, 132, 157)	(119, 120, 169)	(52, 165, 173)	(19, 180, 181)
(57, 176, 185)	(104, 153, 185)	(95, 168, 193)	(28, 195, 197)
(84, 187, 205)	(133, 156, 205)	(21, 220, 221)	(140, 171, 221)
(60, 221, 229)	(105, 208, 233)	(120, 209, 241)	(32, 255, 257)
(23, 264, 265)	(96, 247, 265)	(69, 260, 269)	(115, 252, 277)
(160, 231, 281)	(161, 240, 289)	(68, 285, 293)	

Generating a Triple

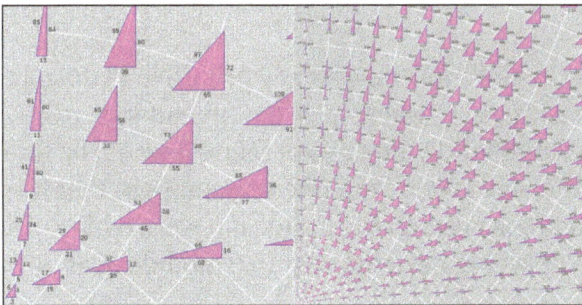

The primitive Pythagorean triples. The odd leg a is plotted on the horizontal axis, the even leg b on the vertical.

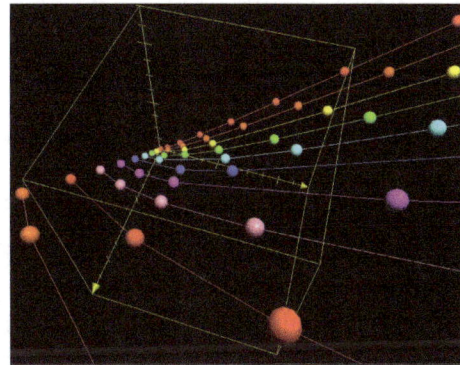

A plot of triples generated by Euclid's formula maps out part of the $z^2 = x^2 + y^2$ cone. A constant m or n traces out part of a parabola on the cone.

Euclid's formula is a fundamental formula for generating Pythagorean triples given an arbitrary pair of integers m and n with $m > n > 0$. The formula states that the integers:

$$a = m^2 - n^2, \ b = 2mn, \ c = m^2 + n^2$$

form a Pythagorean triple. The triple generated by Euclid's formula is primitive if and only if m and n are coprime and not both odd. When both m and n are odd, then a, b, and c will be even, and the triple will not be primitive; however, dividing a, b, and c by 2 will yield a primitive triple when m and n are coprime and both odd.

Every primitive triple arises (after the exchange of a and b, if a is even) from a *unique pair* of coprime numbers m, n, one of which is even. It follows that there are infinitely many primitive Pythagorean triples. This relationship of a, b and c to m and n from Euclid's formula is referenced throughout the rest of this topic.

Despite generating all primitive triples, Euclid's formula does not produce all triples—for example, (9, 12, 15) cannot be generated using integer m and n. This can be remedied by inserting an additional parameter k to the formula. The following will generate all Pythagorean triples uniquely:

$$a = k \cdot (m^2 - n^2),\ b = k \cdot (2mn),\ c = k \cdot (m^2 + n^2)$$

where m, n, and k are positive integers with $m > n$, and with m and n coprime and not both odd.

That these formulas generate Pythagorean triples can be verified by expanding $a^2 + b^2$ using elementary algebra and verifying that the result equals c^2. Since every Pythagorean triple can be divided through by some integer k to obtain a primitive triple, every triple can be generated uniquely by using the formula with m and n to generate its primitive counterpart and then multiplying through by k as in the last equation.

Many formulas for generating triples with particular properties have been developed since the time of Euclid.

Proof of Euclid's Formula

That satisfaction of Euclid's formula by a, b, c is sufficient for the triangle to be Pythagorean is apparent from the fact that for positive integers m and n, $m > n$, the a, b, and c given by the formula are all positive integers, and from the fact that:

$$a^2 + b^2 = (m^2 - n^2)^2 + (2mn)^2 = (m^2 + n^2)^2 = c^2.$$

A proof of the *necessity* that a, b, c be expressed by Euclid's formula for any primitive Pythagorean triple is as follows. All such triples can be written as (a, b, c) where $a^2 + b^2 = c^2$ and a, b, c are coprime. Thus a, b, c are pairwise coprime (if a prime number divided two of them, it would be forced also to divide the third one). As a and b are coprime, one is odd, and one may suppose that it is a, by exchanging, if needed, a and b. This implies that b is even and c is odd (if b were odd, c would be even, and c^2 would be a multiple of 4, while $a^2 + b^2$ would be congruent to 2 modulo 4, as an odd square is congruent to 1 modulo 4).

From $a^2 + b^2 = c^2$ we obtain $c^2 - a^2 = b^2$ and hence $(c - a)(c + a) = b^2$. Then $\frac{(c+a)}{b} = \frac{b}{(c-a)}$. Since $\frac{(c+a)}{b}$ is rational, we set it equal to $\frac{m}{n}$ in lowest terms. Thus $\frac{(c-a)}{b} = \frac{n}{m}$, being the reciprocal of $\frac{(c+a)}{b}$. Then solving:

$$\frac{c}{b} + \frac{a}{b} = \frac{m}{n}, \qquad \frac{c}{b} - \frac{a}{b} = \frac{n}{m}$$

for $\frac{c}{b}$ and $\frac{a}{b}$ gives:

$$\frac{c}{b} = \frac{1}{2}\left(\frac{m}{n} + \frac{n}{m}\right) = \frac{m^2 + n^2}{2mn}, \qquad \frac{a}{b} = \frac{1}{2}\left(\frac{m}{n} - \frac{n}{m}\right) = \frac{m^2 - n^2}{2mn}.$$

As $\frac{m}{n}$ is fully reduced, m and n are coprime, and they cannot both be even. If they were both odd, the numerator of $\frac{m^2 - n^2}{2mn}$ would be a multiple of 4 (because an odd square is congruent to 1 modulo 4), and the denominator $2mn$ would not be a multiple of 4. Since 4 would be the minimum possible even factor in the numerator and 2 would be the maximum possible even factor in the denominator, this would imply a to be even despite defining it as odd. Thus one of m and n is odd and the other is even, and the numerators of the two fractions with denominator $2mn$ are odd. Thus these fractions are fully reduced (an odd prime dividing this denominator divides one of m and n but not the other; thus it does not divide $m^2 \pm n^2$). One may thus equate numerators with numerators and denominators with denominators, giving Euclid's formula,

$$a = m^2 - n^2,\ b = 2mn,\ c = m^2 + n^2 \text{ with } m \text{ and } n \text{ coprime and of opposite parities.}$$

A longer but more commonplace proof is given in Maor and Sierpiński. Another proof is given in Diophantine equation § Example of Pythagorean triples, as an instance of a general method that applies to every homogeneous Diophantine equation of degree two.

Interpretation of Parameters in Euclid's Formula

Suppose the sides of a Pythagorean triangle have lengths $m^2 - n^2$, $2mn$, and $m^2 + n^2$, and suppose the angle between the leg of length $m^2 - n^2$ and the hypotenuse of length $m^2 + n^2$ is denoted as β. Then $\tan \frac{\beta}{2} = \frac{n}{m}$ and the full-angle trigonometric values are $\sin \beta = \frac{2mn}{m^2 + n^2}$, $\cos \beta = \frac{m^2 - n^2}{m^2 + n^2}$, and $\tan \beta = \frac{2mn}{m^2 - n^2}$.

A Variant

The following variant of Euclid's formula is sometimes more convenient, as being more symmetric in m and n (same parity condition on m and n).

If m and n are two odd integers such that $m > n$, then:

$$a = mn,\ b = \frac{m^2 - n^2}{2},\ c = \frac{m^2 + n^2}{2}$$

are three integers that form a Pythagorean triple, which is primitive if and only if m and n are coprime. Conversely, every primitive Pythagorean triple arises (after the exchange of a and b, if a is even) from a unique pair $m > n > 0$ of coprime odd integers.

Elementary Properties of Primitive Pythagorean Triples

General Properties

The properties of a primitive Pythagorean triple (a, b, c) with $a < b < c$ (without specifying which of a or b is even and which is odd) include:

- Is always a perfect square. As it is only a necessary condition but not a sufficient one, it can be used in checking if a given triple of numbers is *not* a Pythagorean triple when they fail the test. For example, the triple {6, 12, 18} passes the test that $(c - a)(c - b)/2$ is a perfect square, but it is not a Pythagorean triple.

- When a triple of numbers a, b and c forms a primitive Pythagorean triple, then (c minus the even leg) and one-half of (c minus the odd leg) are both perfect squares; however this is not a sufficient condition, as the numbers {1, 8, 9} pass the perfect squares test but are not a Pythagorean triple since $1^2 + 8^2 \neq 9^2$.

- At most one of a, b, c is a square.

- The area of a Pythagorean triangle cannot be the square or twice the square of a natural number.

- Exactly one of a, b is odd; c is odd.

- Exactly one of a, b is divisible by 3.

- Exactly one of a, b is divisible by 4.

- Exactly one of a, b, c is divisible by 5.

- The largest number that always divides abc is 60.

- All prime factors of c are primes of the form $4n + 1$. Therefore c is of the form $4n + 1$.

- The area ($K = ab/2$) is a congruent number divisible by 6.

- In every Pythagorean triangle, the radius of the incircle and the radii of the three excircles are natural numbers. Specifically, for a primitive triple the radius of the incircle is $r = n(m - n)$, and the radii of the excircles opposite the sides $m^2 - n^2$, $2mn$, and the hypotenuse $m^2 + n^2$ are respectively $m(m - n)$, $n(m + n)$, and $m(m + n)$.

- As for any right triangle, the converse of Thales' theorem says that the diameter of the circumcircle equals the hypotenuse; hence for primitive triples the circumdiameter is $m^2 + n^2$, and the circumradius is half of this and thus is rational but non-integer (since m and n have opposite parity).

- When the area of a Pythagorean triangle is multiplied by the curvatures of its incircle and 3 excircles, the result is four positive integers $w > x > y > z$, respectively. Integers $-w$, x, y, z satisfy Descartes's Circle Equation. Equivalently, the radius of the outer Soddy circle of any right triangle is equal to its semiperimeter. The outer Soddy center is located at D, where $ACBD$ is a rectangle, ACB the right triangle and AB its hypotenuse.

- Only two sides of a primitive Pythagorean triple can be simultaneously prime because by Euclid's formula for generating a primitive Pythagorean triple, one of the legs must be composite and even. However, only one side can be an integer of perfect power $p \geq 2$ because if two sides were integers of perfect powers with equal exponent p it would contradict the fact that there are no integer solutions to the Diophantine equation $x^{2p} \pm y^{2p} = z^2$, with x, y and z being pairwise coprime.

- There are no Pythagorean triangles in which the hypotenuse and one leg are the legs of another Pythagorean triangle; this is one of the equivalent forms of Fermat's right triangle theorem.

- Each primitive Pythagorean triangle has a ratio of area, K, to squared semiperimeter, s, that is unique to itself and is given by:

$$\frac{K}{s^2} = \frac{n(m-n)}{m(m+n)} = 1 - \frac{c}{s}.$$

- No primitive Pythagorean triangle has an integer altitude from the hypotenuse; that is, every primitive Pythagorean triangle is indecomposable.

- The set of all primitive Pythagorean triples forms a rooted ternary tree in a natural way; see Tree of primitive Pythagorean triples.

- Neither of the acute angles of a Pythagorean triangle can be a rational number of degrees. (This follows from Niven's theorem).

Special Cases

In addition, special Pythagorean triples with certain additional properties can be guaranteed to exist:

- Every integer greater than 2 that is not congruent to 2 mod 4 (in other words, every integer greater than 2 which is *not* of the form $4k + 2$) is part of a primitive Pythagorean triple. (If the integer has the form $4k$, one may take $n =1$ and $m = 2k$ in Euclid's formula; if the integer is $2k + 1$, one may take $n = k$ and $m = k + 1$).

- Every integer greater than 2 is part of a primitive or non-primitive Pythagorean triple. For example, the integers 6, 10, 14, and 18 are not part of primitive triples, but are part of the non-primitive triples (6, 8, 10), (14, 48, 50) and (18, 80, 82).

- There exist infinitely many Pythagorean triples in which the hypotenuse and the longest leg differ by exactly one. Such triples are necessarily primitive and have the form $(2n + 1, 2n^2 + 2n, 2n^2 + 2n +1)$. This results from Euclid's formula by remarking that the condition implies that the triple is primitive and must verify $(m^2 + n^2) - 2mn = 1$. This implies $(m - n)^2 = 1$, and thus $m = n + 1$. The above form of the triples results thus of substituting m for $n + 1$ in Euclid's formula.

- There exist infinitely many primitive Pythagorean triples in which the hypotenuse and the longest leg differ by exactly two. They are all primitive, and are obtained by putting $n = 1$ in Euclid's formula. More generally, for every integer $k > 0$, there exist infinitely many primitive Pythagorean triples in which the hypotenuse and the odd leg differ by $2k^2$. They are obtained by putting $n = k$ in Euclid's formula.

- There exist infinitely many Pythagorean triples in which the two legs differ by exactly one. For example, $20^2 + 21^2 = 29^2$; these are generated by Euclid's formula when $\frac{m-n}{n}$ is a convergent to $\sqrt{2}$.

- For each natural number k, there exist k Pythagorean triples with different hypotenuses and the same area.

- For each natural number k, there exist at least k different primitive Pythagorean triples with the same leg a, where a is some natural number (the length of the even leg is $2mn$, and it suffices to choose a with many factorizations, for example $a = 4b$, where b is a product of k different odd primes; this produces at least 2^k different primitive triples).

- For each natural number n, there exist at least n different Pythagorean triples with the same hypotenuse.

- There exist infinitely many Pythagorean triples with square numbers for both the hypotenuse c and the sum of the legs $a + b$. According to Fermat, the smallest such triple has sides $a = 4{,}565{,}486{,}027{,}761$; $b = 1{,}061{,}652{,}293{,}520$; and $c = 4{,}687{,}298{,}610{,}289$. Here $a + b = 2{,}372{,}159^2$ and $c = 2{,}165{,}017^2$. This is generated by Euclid's formula with parameter values $m = 2{,}150{,}905$ and $n = 246{,}792$.

- There exist non-primitive Pythagorean triangles with integer altitude from the hypotenuse. Such Pythagorean triangles are known as decomposable since they can be split along this altitude into two separate and smaller Pythagorean triangles.

Geometry of Euclid's Formula

Rational Points on a Unit Circle

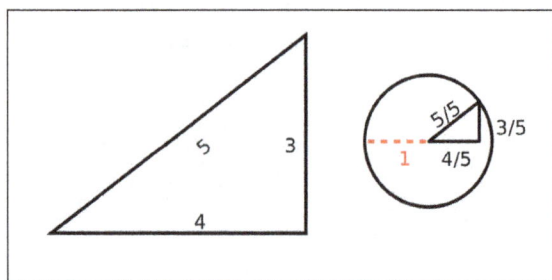

3,4,5 maps to x,y point (4/5,3/5) on the unit circle.

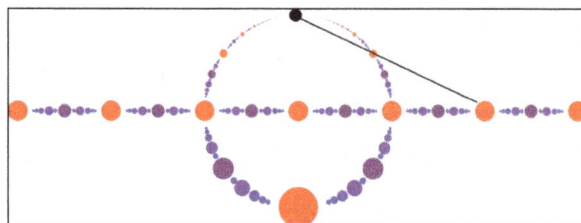

The rational points on a circle correspond, under stereographic projection, to the rational points of the line.

Euclid's formula for a Pythagorean triple:

$$a = 2mn, \quad b = m^2 - n^2, \quad c = m^2 + n^2$$

can be understood in terms of the geometry of rational points on the unit circle.

In fact, a point in the Cartesian plane with coordinates (x, y) belongs to the unit circle if $x^2 + y^2 = 1$. The point is *rational* if x and y are rational numbers, that is, if there are coprime integers a, b, c such that:

$$\left(\frac{a}{c}\right)^2 + \left(\frac{b}{c}\right)^2 = 1.$$

By multiplying both members by c^2, one can see that the rational points on the circle are in one-to-one correspondence with the primitive Pythagorean triples.

The unit circle may also be defined by a parametric equation:

$$x = \frac{1 - t^2}{1 + t^2} \quad y = \frac{2t}{1 + t^2}.$$

Euclid's formula for Pythagorean triples means that, except for $(-1, 0)$, a point on the circle is rational if and only if the corresponding value of t is a rational number.

Stereographic Approach

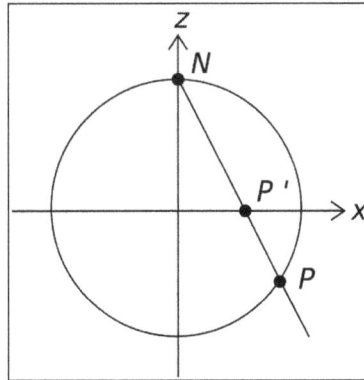

Stereographic projection of the unit circle onto the x-axis. Given a point P on the unit circle, draw a line from P to the point $N = (0, 1)$ (the *north pole*). The point P' where the line intersects the x-axis is the stereographic projection of P. Inversely, starting with a point P' on the x-axis, and drawing a line from P' to N, the inverse stereographic projection is the point P where the line intersects the unit circle.

There is a correspondence between points on the unit circle with rational coordinates and primitive Pythagorean triples. At this point, Euclid's formulae can be derived either by methods of trigonometry or equivalently by using the stereographic projection.

For the stereographic approach, suppose that P' is a point on the x-axis with rational coordinates

$$P' = \left(\frac{m}{n}, 0 \right).$$

Then, it can be shown by basic algebra that the point P has coordinates

$$P = \left(\frac{2\left(\frac{m}{n}\right)}{\left(\frac{m}{n}\right)^2 + 1}, \frac{\left(\frac{m}{n}\right)^2 - 1}{\left(\frac{m}{n}\right)^2 + 1} \right) = \left(\frac{2mn}{m^2 + n^2}, \frac{m^2 - n^2}{m^2 + n^2} \right).$$

This establishes that each rational point of the x-axis goes over to a rational point of the unit circle. The converse, that every rational point of the unit circle comes from such a point of the x-axis, follows by applying the inverse stereographic projection. Suppose that $P(x, y)$ is a point of the unit circle with x and y rational numbers. Then the point P' obtained by stereographic projection onto the x-axis has coordinates

$$\left(\frac{x}{1 - y}, 0 \right)$$

which is rational.

In terms of algebraic geometry, the algebraic variety of rational points on the unit circle is birational to the affine line over the rational numbers. The unit circle is thus called a rational curve, and it is this fact which enables an explicit parameterization of the (rational number) points on it by means of rational functions.

Pythagorean Triangles in a 2D Lattice

A 2D lattice is a regular array of isolated points where if any one point is chosen as the Cartesian origin $(0, 0)$, then all the other points are at (x, y) where x and y range over all positive and negative integers. Any Pythagorean triangle with triple (a, b, c) can be drawn within a 2D lattice with vertices at coordinates $(0, 0)$, $(a, 0)$ and $(0, b)$. The count of lattice points lying strictly within the bounds of the triangle is given by $\dfrac{(a-1)(b-1)-\gcd(a,b)+1}{2}$, for primitive Pythagorean triples this interior lattice count is $\dfrac{(a-1)(b-1)}{2}$. The area (by Pick's theorem equal to one less than the interior lattice count plus half the boundary lattice count) equals $\dfrac{ab}{2}$.

The first occurrence of two primitive Pythagorean triples sharing the same area occurs with triangles with sides $(20, 21, 29)$, $(12, 35, 37)$ and common area 210 (sequence A093536 in the OEIS). The first occurrence of two primitive Pythagorean triples sharing the same interior lattice count occurs with $(18108, 252685, 253333)$, $(28077, 162964, 165365)$ and interior lattice count 2287674594. Three primitive Pythagorean triples have been found sharing the same area: $(4485, 5852, 7373)$, $(3059, 8580, 9109)$, $(1380, 19019, 19069)$ with area 13123110. As yet, no set of three primitive Pythagorean triples have been found sharing the same interior lattice count.

Enumeration of Primitive Pythagorean Triples

By Euclid's formula all primitive Pythagorean triples can be generated from integers m and n with $m > n > 0$, $m+n$ odd and $\gcd(m,n)=1$.

Hence there is a 1 to 1 mapping of rationals (in lowest terms) to primitive Pythagorean triples where $\dfrac{n}{m}$ is in the interval $(0,1)$ and $m+n$ odd.

The reverse mapping from a primitive triple (a,b,c) where $c > b > a > 0$ to a rational $\dfrac{n}{m}$ is achieved by studying the two sums $a+c$ and $b+c$. One of these sums will be a square that can be equated to $(m+n)^2$ and the other will be twice a square that can be equated to $2m^2$. It is then possible to determine the rational $\dfrac{n}{m}$.

In order to enumerate primitive Pythagorean triples the rational can be expressed as an ordered pair (n,m) and mapped to an integer using a pairing function such as Cantor's pairing function:

$8,18,19,32,33,34,\ldots$ and gives rationals.

$\dfrac{1}{2}, \dfrac{2}{3}, \dfrac{1}{4}, \dfrac{3}{4}, \dfrac{2}{5}, \dfrac{1}{6}, \ldots$ these, in turn, generate primitive triples.

$(3,4,5),(5,12,13),(8,15,17),(7,24,25),(20,21,29),(12,35,37),\ldots$

Spinors and the Modular Group

Pythagorean triples can likewise be encoded into a matrix of the form:

$$X = \begin{bmatrix} c+b & a \\ a & c-b \end{bmatrix}.$$

A matrix of this form is symmetric. Furthermore, the determinant of X is:

$$\det X = c^2 - a^2 - b^2$$

which is zero precisely when (a,b,c) is a Pythagorean triple. If X corresponds to a Pythagorean triple, then as a matrix it must have rank 1.

Since X is symmetric, it follows from a result in linear algebra that there is a column vector $\xi = [m\ n]^T$ such that the outer product:

$$X = 2\begin{bmatrix} m \\ n \end{bmatrix}[m\ n] = 2\xi\xi^T$$

holds, where the T denotes the matrix transpose. The vector ξ is called a spinor (for the Lorentz group $SO(1, 2)$). In abstract terms, the Euclid formula means that each primitive Pythagorean triple can be written as the outer product with itself of a spinor with integer entries, as in above equation.

The modular group Γ is the set of 2×2 matrices with integer entries:

$$A = \begin{bmatrix} \alpha & \beta \\ \gamma & \delta \end{bmatrix}$$

with determinant equal to one: $\alpha\delta - \beta\gamma = 1$. This set forms a group, since the inverse of a matrix in Γ is again in Γ, as is the product of two matrices in Γ. The modular group acts on the collection of all integer spinors. Furthermore, the group is transitive on the collection of integer spinors with relatively prime entries. For if $[m\ n]^T$ has relatively prime entries, then:

$$\begin{bmatrix} m & -v \\ n & u \end{bmatrix}\begin{bmatrix} 1 \\ 0 \end{bmatrix} = \begin{bmatrix} m \\ n \end{bmatrix}$$

where u and v are selected (by the Euclidean algorithm) so that $mu + nv = 1$.

By acting on the spinor ξ in $X = 2\begin{bmatrix} m \\ n \end{bmatrix}[m\ n] = 2\xi\xi^T$, the action of Γ goes over to an action on Pythagorean triples, provided one allows for triples with possibly negative components. Thus if A is a matrix in Γ, then:

$$2(A\xi)(A\xi)^T = AXA^T$$

gives rise to an action on the matrix X in $X = 2\begin{bmatrix} m \\ n \end{bmatrix}[m\ n] = 2\xi\xi^T$. This does not give a well-defined action on primitive triples, since it may take a primitive triple to an imprimitive one. It is convenient at this point to call a triple (a,b,c) standard if $c > 0$ and either (a,b,c) are relatively prime or

$(a/2,b/2,c/2)$ are relatively prime with $a/2$ odd. If the spinor $[m\ n]^T$ has relatively prime entries, then the associated triple (a,b,c) determined by $X = 2\begin{bmatrix} m \\ n \end{bmatrix}[m\ n] = 2\xi\xi^T$ is a standard triple. It follows that the action of the modular group is transitive on the set of standard triples.

Alternatively, restrict attention to those values of m and n for which m is odd and n is even. Let the subgroup $\Gamma(2)$ of Γ be the kernel of the group homomorphism:

$$\Gamma = SL(2,\mathbf{Z}) \rightarrow SL(2,\mathbf{Z}_2)$$

where $SL(2,\mathbf{Z}_2)$ is the special linear group over the finite field $\mathbf{Z}_2$ of integers modulo 2. Then $\Gamma(2)$ is the group of unimodular transformations which preserve the parity of each entry. Thus if the first entry of ξ is odd and the second entry is even, then the same is true of $A\xi$ for all $A \in \Gamma(2)$. In fact, under the action $2(A\xi)(A\xi)^T = AXA^T$, the group $\Gamma(2)$ acts transitively on the collection of primitive Pythagorean triples.

The group $\Gamma(2)$ is the free group whose generators are the matrices:

$$U = \begin{bmatrix} 1 & 2 \\ 0 & 1 \end{bmatrix}, \qquad L = \begin{bmatrix} 1 & 0 \\ 2 & 1 \end{bmatrix}.$$

Consequently, every primitive Pythagorean triple can be obtained in a unique way as a product of copies of the matrices U and L.

Parent and Child Relationships

By a result of Berggren, all primitive Pythagorean triples can be generated from the (3, 4, 5) triangle by using the three linear transformations T_1, T_2, T_3 below, where a, b, c are sides of a triple:

	new side a	new side b	new side c
T_1:	$a - 2b + 2c$	$2a - b + 2c$	$2a - 2b + 3c$
T_2:	$a + 2b + 2c$	$2a + b + 2c$	$2a + 2b + 3c$
T_3:	$-a + 2b + 2c$	$-2a + b + 2c$	$-2a + 2b + 3c$

In other words, every primitive triple will be a "parent" to three additional primitive triples. Starting from the initial node with $a = 3$, $b = 4$, and $c = 5$, the operation T_1 produces the new triple:

$$(3 - (2\times4) + (2\times5), (2\times3) - 4 + (2\times5), (2\times3) - (2\times4) + (3\times5)) = (5, 12, 13),$$

and similarly T_2 and T_3 produce the triples (21, 20, 29) and (15, 8, 17).

The linear transformations T_1, T_2, and T_3 have a geometric interpretation in the language of quadratic forms. They are closely related to (but are not equal to) reflections generating the orthogonal group of $x^2 + y^2 - z^2$ over the integers.

Relation to Gaussian Integers

Alternatively, Euclid's formulae can be analyzed and proven using the Gaussian integers. Gaussian integers are complex numbers of the form $\alpha = u + vi$, where u and v are ordinary integers and i is the square root of negative one. The units of Gaussian integers are ±1 and ±i. The ordinary integers are called the rational integers and denoted as Z. The Gaussian integers are denoted as Z[i]. The right-hand side of the Pythagorean theorem may be factored in Gaussian integers:

$$c^2 = a^2 + b^2 = (a + bi)\overline{(a + bi)} = (a + bi)(a - bi).$$

A primitive Pythagorean triple is one in which a and b are coprime, i.e., they share no prime factors in the integers. For such a triple, either a or b is even, and the other is odd; from this, it follows that c is also odd.

The two factors $z := a + bi$ and $z^* := a - bi$ of a primitive Pythagorean triple each equal the square of a Gaussian integer. This can be proved using the property that every Gaussian integer can be factored uniquely into Gaussian primes up to units. This unique factorization follows from the fact that, roughly speaking, a version of the Euclidean algorithm can be defined on them. The proof has three steps. First, if a and b share no prime factors in the integers, then they also share no prime factors in the Gaussian integers. Assume $a = gu$ and $b = gv$ with Gaussian integers g, u and v and g not a unit. Then u and v lie on the same line through the origin. All Gaussian integers on such a line are integer multiples of some Gaussian integer h. But then the integer $gh \neq \pm1$ divides both a and b. Second, it follows that z and z^* likewise share no prime factors in the Gaussian integers. For if they did, then their common divisor δ would also divide $z + z^* = 2a$ and $z - z^* = 2ib$. Since a and b are coprime, that implies that δ divides 2 = $(1 + i)(1 - i) = i(1 - i)^2$. From the formula $c^2 = zz^*$, that in turn would imply that c is even, contrary to the hypothesis of a primitive Pythagorean triple. Third, since c^2 is a square, every Gaussian prime in its factorization is doubled, i.e., appears an even number of times. Since z and z^* share no prime factors, this doubling is also true for them. Hence, z and z^* are squares.

Thus, the first factor can be written:

$$a + bi = \varepsilon\left(m + ni\right)^2, \quad \varepsilon \in \{\pm1, \pm i\}.$$

The real and imaginary parts of this equation give the two formulas:

$$\begin{cases} \varepsilon = +1, & a = +\left(m^2 - n^2\right), \quad b = +2mn; \\ \varepsilon = -1, & a = -\left(m^2 - n^2\right), \quad b = -2mn; \\ \varepsilon = +i, & a = -2mn, \quad b = +\left(m^2 - n^2\right); \\ \varepsilon = -i, & a = +2mn, \quad b = -\left(m^2 - n^2\right). \end{cases}$$

For any primitive Pythagorean triple, there must be integers m and n such that these two equations are satisfied. Hence, every Pythagorean triple can be generated from some choice of these integers.

As Perfect Square Gaussian Integers

If we consider the square of a Gaussian integer we get the following direct interpretation of Euclid's formulae as representing a perfect square Gaussian integers.

$$(m + ni)^2 = (m^2 - n^2) + 2mni.$$

Using the facts that the Gaussian integers are a Euclidean domain and that for a Gaussian integer p $|p|^2$ is always a square it is possible to show that a Pythagorean triples correspond to the square of a prime Gaussian integer if the hypotenuse is prime.

If the Gaussian integer is not prime then it is the product of two Gaussian integers p and q with $|p|^2$ and $|q|^2$ integers. Since magnitudes multiply in the Gaussian integers, the product must be $p\|q\|$, which when squared to find a Pythagorean triple must be composite. The contrapositive completes the proof.

Relation to Ellipses with Integral Dimensions

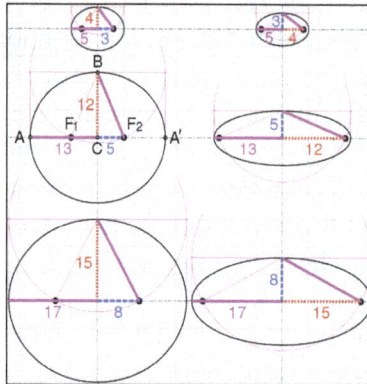

Relationship between Pythagorean triples and ellipses with integral linear eccentricity, and major and minor axes, for the first 3 Pythagorean triples.

With reference to the figure and the definition of the foci of an ellipse, F_1 and F_2, for any point P on the ellipse, $F_1P + PF_2$ is constant.

As points A and B are both on the ellipse, $F_1A + AF_2 = F_1B + BF_2$. Due to symmetry, $F_1A + AF_2 = F_2A' + AF_2 = AA' = 2\,AC$, and $F_1B + BF_2 = 2\,BF_2$. Hence, $AC = BF_2$.

Thus, if BCF_2 is a right-angle triangle with integral sides, the separation of the foci, linear eccentricity, minor axis and major axis are all also integers.

Distribution of Triples

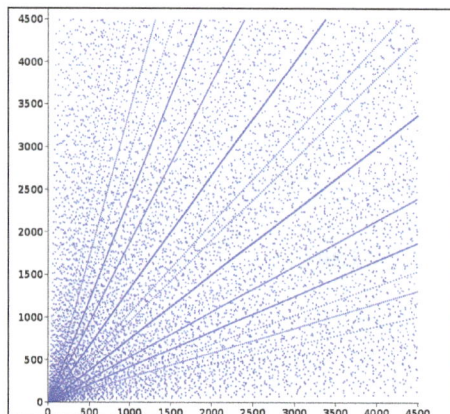

A scatter plot of the legs (a,b) of the first Pythagorean triples with a and b less than 4500.

There are a number of results on the distribution of Pythagorean triples. In the scatter plot, a number of obvious patterns are already apparent. Whenever the legs (a,b) of a primitive triple appear in the plot, all integer multiples of (a,b) must also appear in the plot, and this property produces the appearance of lines radiating from the origin in the diagram.

Within the scatter, there are sets of parabolic patterns with a high density of points and all their foci at the origin, opening up in all four directions. Different parabolas intersect at the axes and appear to reflect off the axis with an incidence angle of 45 degrees, with a third parabola entering in a perpendicular fashion. Within this quadrant, each arc centered on the origin shows that section of the parabola that lies between its tip and its intersection with its semi-latus rectum.

These patterns can be explained as follows - If $a^2 / 4n$ is an integer, then $(a, |n - a^2 / 4n|, n + a^2 / 4n)$ is a Pythagorean triple. In fact every Pythagorean triple (a, b, c) can be written in this way with integer n, possibly after exchanging a and b, since $n = (b + c) / 2$ and a and b cannot both be odd. The Pythagorean triples thus lie on curves given by $b = |n - a^2 / 4n|$, that is, parabolas reflected at the a-axis, and the corresponding curves with a and b interchanged. If a is varied for a given n (i.e. on a given parabola), integer values of b occur relatively frequently if n is a square or a small multiple of a square. If several such values happen to lie close together, the corresponding parabolas approximately coincide, and the triples cluster in a narrow parabolic strip. For instance, $38^2 = 1444$, $2 \times 27^2 = 1458$, $3 \times 22^2 = 1452$, $5 \times 17^2 = 1445$ and $10 \times 12^2 = 1440$; the corresponding parabolic strip around $n \approx 1450$ is clearly visible in the scatter plot.

The angular properties described above follow immediately from the functional form of the parabolas. The parabolas are reflected at the a-axis at $a = 2n$, and the derivative of b with respect to a at this point is -1; hence the incidence angle is $45°$. Since the clusters, like all triples, are repeated at integer multiples, the value $2n$ also corresponds to a cluster. The corresponding parabola intersects the b-axis at right angles at $b = 2n$, and hence its reflection upon interchange of a and b intersects the a-axis at right angles at $a = 2n$, precisely where the parabola for n is reflected at the a-axis. The same is of course true for a and b interchanged.

Special Cases and Related Equations

The Platonic Sequence

The case $n = 1$ of the more general construction of Pythagorean triples has been known for a long time. Proclus, in his commentary to the 47th Proposition of the first book of Euclid's Elements, describes it as follows:

> Certain methods for the discovery of triangles of this kind are handed down, one which they refer to Plato, and another to Pythagoras. (The latter) starts from odd numbers. For it makes the odd number the smaller of the sides about the right angle; then it takes the square of it, subtracts unity and makes half the difference the greater of the sides about the right angle; lastly it adds unity to this and so forms the remaining side, the hypotenuse.

For the method of Plato argues from even numbers. It takes the given even number and makes it one of the sides about the right angle; then, bisecting this number and squaring the half, it adds unity to the square to form the hypotenuse, and subtracts unity from the square to form the other side about the right angle. Thus it has formed the same triangle that which was obtained by the other method.

In equation form, this becomes:

a is odd:

$$\text{side } a : \text{side } b = \frac{a^2 - 1}{2} : \text{side } c = \frac{a^2 + 1}{2}.$$

a is even:

$$\text{side } a : \text{side } b = \left(\frac{a}{2}\right)^2 - 1 : \text{side } c = \left(\frac{a}{2}\right)^2 + 1.$$

It can be shown that all Pythagorean triples can be obtained, with appropriate rescaling, from the basic Platonic sequence $(a, (a^2 - 1)/2$ and $(a^2 + 1)/2)$ by allowing a to take non-integer rational values. If a is replaced with the fraction m/n in the sequence, the result is equal to the 'standard' triple generator $(2mn, m^2 - n^2, m^2 + n^2)$ after rescaling. It follows that every triple has a corresponding rational a value which can be used to generate a similar triangle (one with the same three angles and with sides in the same proportions as the original). For example, the Platonic equivalent of (56, 33, 65) is generated by $a = m/n = 7/4$ as $(a, (a^2 - 1)/2, (a^2 + 1)/2) = (56/32, 33/32, 65/32)$. The Platonic sequence itself can be derived by following the steps for 'splitting the square.

The Jacobi–Madden Equation

The equation,

$$a^4 + b^4 + c^4 + d^4 = (a + b + c + d)^4$$

Is equivalent to the special Pythagorean triple,

$$(a^2 + ab + b^2)^2 + (c^2 + cd + d^2)^2 = ((a + b)^2 + (a + b)(c + d) + (c + d)^2)^2$$

There is an infinite number of solutions to this equation as solving for the variables involves an elliptic curve. Small ones are,

$$a, b, c, d = -2634, 955, 1770, 5400$$

$$a, b, c, d = -31764, 7590, 27385, 48150$$

Equal Sums of Two Squares

One way to generate solutions to $a^2 + b^2 = c^2 + d^2$ is to parametrize a, b, c, d in terms of integers m, n, p, q as follows:

$$(m^2 + n^2)(p^2 + q^2) = (mp - nq)^2 + (np + mq)^2 = (mp + nq)^2 + (np - mq)^2.$$

Equal Sums of Two Fourth Powers

Given two sets of Pythagorean triples,

$$(a^2 - b^2)^2 + (2ab)^2 = (a^2 + b^2)^2$$

$$(c^2 - d^2)^2 + (2cd)^2 = (c^2 + d^2)^2$$

the problem of finding equal products of a non-hypotenuse side and the hypotenuse,

$$(a^2 - b^2)(a^2 + b^2) = (c^2 - d^2)(c^2 + d^2)$$

is easily seen to be equivalent to the equation,

$$a^4 - b^4 = c^4 - d^4$$

and was first solved by Euler as $a,b,c,d = 133,59,158,134$. Since he showed this is a rational point in an elliptic curve, then there is an infinite number of solutions. In fact, he also found a 7th degree polynomial parameterization.

Descartes' Circle Theorem

For the case of Descartes' circle theorem where all variables are squares,

$$2(a^4 + b^4 + c^4 + d^4) = (a^2 + b^2 + c^2 + d^2)^2$$

Euler showed this is equivalent to three simultaneous Pythagorean triples,

$$(2ab)^2 + (2cd)^2 = (a^2 + b^2 - c^2 - d^2)^2$$

$$(2ac)^2 + (2bd)^2 = (a^2 - b^2 + c^2 - d^2)^2$$

$$(2ad)^2 + (2bc)^2 = (a^2 - b^2 - c^2 + d^2)^2$$

There is also an infinite number of solutions, and for the special case when $a + b = c$, then the equation simplifies to,

$$4(a^2 + ab + b^2) = d^2$$

with small solutions as $a,b,c,d = 3,5,8,14$ and can be solved as binary quadratic forms.

Almost-isosceles Pythagorean Triples

No Pythagorean triples are isosceles, because the ratio of the hypotenuse to either other side is $\sqrt{2}$, but $\sqrt{2}$ cannot be expressed as the ratio of 2 integers.

There are, however, right-angled triangles with integral sides for which the lengths of the non-hypotenuse sides differ by one, such as,

$$3^2 + 4^2 = 5^2$$

$$20^2 + 21^2 = 29^2$$

and an infinite number of others. They can be completely parameterized as,

$$\left(\tfrac{x-1}{2}\right)^2 + \left(\tfrac{x+1}{2}\right)^2 = y^2$$

where $\{x, y\}$ are the solutions to the Pell equation $x^2 - 2y^2 = -1$.

If a, b, c are the sides of this type of primitive Pythagorean triple (PPT) then the solution to the Pell equation is given by the recursive formula:

$$a_n = 6a_{n-1} - a_{n-2} + 2 \text{ with } a_1 = 3 \text{ and } a_2 = 20$$

$$b_n = 6b_{n-1} - b_{n-2} - 2 \text{ with } b_1 = 4 \text{ and } b_2 = 21$$

$$c_n = 6c_{n-1} - c_{n-2} \text{ with } c_1 = 5 \text{ and } c_2 = 29.$$

This sequence of PPTs forms the central stem (trunk) of the rooted ternary tree of PPTs.

When it is the longer non-hypotenuse side and hypotenuse that differ by one, such as in:

$$5^2 + 12^2 = 13^2$$

$$7^2 + 24^2 = 25^2$$

then the complete solution for the PPT a, b, c is:

$$a = 2m + 1, \quad b = 2m^2 + 2m, \quad c = 2m^2 + 2m + 1$$

and

$$(2m + 1)^2 + (2m^2 + 2m)^2 = (2m^2 + 2m + 1)^2$$

where integer $m > 0$ is the generating parameter.

It shows that all odd numbers (greater than 1) appear in this type of almost-isosceles PPT. This sequence of PPTs forms the right hand side outer stem of the rooted ternary tree of PPTs.

Another property of this type of almost-isosceles PPT is that the sides are related such that:

$$a^b + b^a = Kc$$

for some integer K. Or in other words $a^b + b^a$ is divisible by c such as in:

$$(5^{12} + 12^5)/13 = 18799189.$$

Fibonacci Numbers in Pythagorean Triples

Starting with 5, every second Fibonacci number is the length of the hypotenuse of a right triangle with integer sides, or in other words, the largest number in a Pythagorean triple. The length of the longer leg of this triangle is equal to the sum of the three sides of the preceding triangle in this series of triangles, and the shorter leg is equal to the difference between the preceding bypassed Fibonacci number and the shorter leg of the preceding triangle.

Generalizations

There are several ways to generalize the concept of Pythagorean triples.

Pythagorean Quadruple

A set of four positive integers a, b, c and d such that $a^2 + b^2 + c^2 = d^2$ is called a Pythagorean quadruple. The simplest example is $(1, 2, 2, 3)$, since $1^2 + 2^2 + 2^2 = 3^2$. The next simplest (primitive) example is $(2, 3, 6, 7)$, since $2^2 + 3^2 + 6^2 = 7^2$.

All quadruples are given by the formula:

$$(m^2 + n^2 - p^2 - q^2)^2 + (2mq + 2np)^2 + (2nq - 2mp)^2 = (m^2 + n^2 + p^2 + q^2)^2$$

Pythagorean n-tuple

Using the simple algebraic identity,

$$(x_1^2 - x_0)^2 + (2x_1)^2 x_0 = (x_1^2 + x_0)^2$$

for arbitrary x_0, x_1, it is easy to prove that the square of the sum of n squares is itself the sum of n squares by letting $x_0 = x_2^2 + x_3^2 + \ldots + x_n^2$ and then distributing terms. One can see how Pythagorean triples and quadruples are just the particular cases $x_0 = x_2^2$ and $x_0 = x_2^2 + x_3^2$, respectively, and so on for other n, with quintuples given by:

$$(a^2 - b^2 - c^2 - d^2)^2 + (2ab)^2 + (2ac)^2 + (2ad)^2 = (a^2 + b^2 + c^2 + d^2)^2.$$

Since the sum $F(k,m)$ of k consecutive squares beginning with m^2 is given by the formula,

$$F(k,m) = km(k - 1 + m) + \frac{k(k-1)(2k-1)}{6}$$

one may find values (k, m) so that $F(k,m)$ is a square, such as one by Hirschhorn where the number of terms is itself a square,

$$m = \frac{v^4 - 24v^2 - 25}{48}, \; k = v^2, F(m,k) = \frac{v^5 + 47v}{48}$$

and $v \geq 5$ is any integer not divisible by 2 or 3. For the smallest case $v = 5$, hence $k = 25$, this yields the well-known cannonball-stacking problem of Lucas,

$$0^2 + 1^2 + 2^2 + \ldots + 24^2 = 70^2$$

a fact which is connected to the Leech lattice.

In addition, if in a Pythagorean n-tuple ($n \geq 4$) all addends are consecutive except one, one can use the equation,

$$F(k,m) + p^2 = (p+1)^2$$

Since the second power of p cancels out, this is only linear and easily solved for as $p = \dfrac{F(k,m) - 1}{2}$

though k, m should be chosen so that p is an integer, with a small example being $k = 5$, $m = 1$ yielding,

$$1^2 + 2^2 + 3^2 + 4^2 + 5^2 + 27^2 = 28^2$$

Thus, one way of generating Pythagorean n-tuples is by using, for various x,

$$x^2 + (x+1)^2 + \cdots + (x+q)^2 + p^2 = (p+1)^2,$$

where $q = n-2$ and where:

$$p = \frac{(q+1)x^2 + q(q+1)x + \dfrac{q(q+1)(2q+1)}{6} - 1}{2}.$$

Fermat's Last Theorem

A generalization of the concept of Pythagorean triples is the search for triples of positive integers a, b, and c, such that $a^n + b^n = c^n$, for some n strictly greater than 2. Pierre de Fermat in 1637 claimed that no such triple exists, a claim that came to be known as Fermat's Last Theorem because it took longer than any other conjecture by Fermat to be proven or disproven. The first proof was given by Andrew Wiles in 1994.

$n - 1$ or n nth Powers Summing to an nth Power

Another generalization is searching for sequences of $n + 1$ positive integers for which the nth power of the last is the sum of the nth powers of the previous terms. The smallest sequences for known values of n are:

- $n = 3$: {3, 4, 5; 6}

- $n = 4$: {30, 120, 272, 315; 353}

- $n = 5$: {19, 43, 46, 47, 67; 72}

- $n = 7$: {127, 258, 266, 413, 430, 439, 525; 568}

- $n = 8$: {90, 223, 478, 524, 748, 1088, 1190, 1324; 1409}

For the $n=3$ case, in which $x^3 + y^3 + z^3 = w^3$, called the Fermat cubic, a general formula exists giving all solutions.

A slightly different generalization allows the sum of $(k + 1)$ nth powers to equal the sum of $(n - k)$ nth powers. For example:

($n = 3$): $1^3 + 12^3 = 9^3 + 10^3$, made famous by Hardy's recollection of a conversation with Ramanujan about the number 1729 being the smallest number that can be expressed as a sum of two cubes in two distinct ways.

There can also exist $n - 1$ positive integers whose nth powers sum to an nth power (though, by Fermat's last theorem, not for $n = 3$); these are counterexamples to Euler's sum of powers conjecture. The smallest known counterexamples are

- $n = 4$: (95800, 217519, 414560; 422481)
- $n = 5$: (27, 84, 110, 133; 144)

Heronian Triangle Triples

A Heronian triangle is commonly defined as one with integer sides whose area is also an integer, and we shall consider Heronian triangles with *distinct* integer sides. The lengths of the sides of such a triangle form a Heronian triple (a, b, c) provided $a < b < c$. Clearly, any Pythagorean triple is a Heronian triple, since in a Pythagorean triple at least one of the legs a, b must be even, so that the area $ab/2$ is an integer. Not every Heronian triple is a Pythagorean triple, however, as the example (4, 13, 15) with area 24 shows.

If (a, b, c) is a Heronian triple, so is (ma, mb, mc) where m is any positive integer greater than one. The Heronian triple (a, b, c) is primitive provided a, b, c are pairwise relatively prime (as with a Pythagorean triple). Here are a few of the simplest primitive Heronian triples that are not Pythagorean triples:

(4, 13, 15) with area 24

(3, 25, 26) with area 36

(7, 15, 20) with area 42

(6, 25, 29) with area 60

(11, 13, 20) with area 66

(13, 14, 15) with area 84

(13, 20, 21) with area 126

By Heron's formula, the extra condition for a triple of positive integers (a, b, c) with $a < b < c$ to be Heronian is that:

$$(a^2 + b^2 + c^2)^2 - 2(a^4 + b^4 + c^4)$$

or equivalently:

$$2(a^2b^2 + a^2c^2 + b^2c^2) - (a^4 + b^4 + c^4)$$

be a nonzero perfect square divisible by 16.

Application to Cryptography

Primitive Pythagorean triples have been used in cryptography as random sequences and for the generation of keys.

Pythagorean Theorem

Pythagorean theorem is the well-known geometric theorem that the sum of the squares on the legs of a right triangle is equal to the square on the hypotenuse (the side opposite the right angle)—or, in familiar algebraic notation, $a^2 + b^2 = c^2$. Although the theorem has long been associated with Greek mathematician-philosopher Pythagoras, it is actually far older. Four Babylonian tablets from circa 1900–1600 BCE indicate some knowledge of the theorem, or at least of special integers known as Pythagorean triples that satisfy it. The theorem is mentioned in the Baudhayana Sulba-sutra of India, which was written between 800 and 400 BCE. Nevertheless, the theorem came to be credited to Pythagoras. It is also proposition number 47 from Book I of Euclid's Elements.

According to the Syrian historian Iamblichus, Pythagoras was introduced to mathematics by Thales of Miletus and his pupil Anaximander. In any case, it is known that Pythagoras traveled to Egypt about 535 BCE to further his study, was captured during an invasion in 525 BCE by Cambyses II of Persia and taken to Babylon, and may possibly have visited India before returning to the Mediterranean. Pythagoras soon settled in Croton (now Crotone, Italy) and set up a school, or in modern terms a monastery, where all members took strict vows of secrecy, and all new mathematical results for several centuries were attributed to his name. Thus, not only is the first proof of the theorem not known, there is also some doubt that Pythagoras himself actually proved the theorem that bears his name. Some scholars suggest that the first proof was the one shown in the figure. It was probably independently discovered in several different cultures.

Book I of the Elements ends with Euclid's famous "windmill" proof of the Pythagorean theorem. Later in Book VI of the Elements, Euclid delivers an even easier demonstration using the proposition that the areas of similar triangles are proportionate to the squares of their corresponding sides. Apparently, Euclid invented the windmill proof so that he could place the Pythagorean theorem as the capstone to Book I. He had not yet demonstrated (as he would in Book V) that line lengths can be manipulated in proportions as if they were commensurable numbers (integers or ratios of integers). The problem he faced is explained in the Sidebar: Incommensurables.

A great many different proofs and extensions of the Pythagorean theorem have been invented. Taking extensions first, Euclid himself showed in a theorem praised in antiquity that any symmetrical regular figures drawn on the sides of a right triangle satisfy the Pythagorean relationship: the figure drawn on the hypotenuse has an area equal to the sum of the areas of the figures drawn on the legs. The semicircles that define Hippocrates of Chios's lunes are examples of such an extension.

In the Nine Chapters on the Mathematical Procedures (or Nine Chapters), compiled in the 1st century CE in China, several problems are given, along with their solutions, that involve finding the length of one of the sides of a right triangle when given the other two sides. In the Commentary of Liu Hui, from the 3rd century, Liu Hui offered a proof of the Pythagorean theorem that called for cutting up the squares on the legs of the right triangle and rearranging them ("tangram style") to correspond to the square on the hypotenuse. Although his original drawing does not survive, the next figure shows a possible reconstruction.

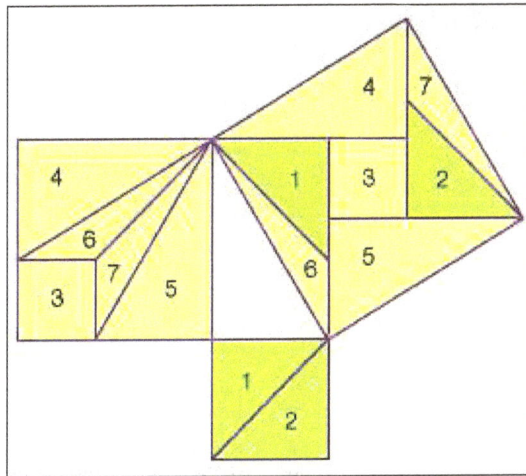

"Tangram" proof of the Pythagorean theorem by Liu Hui, 3rd century AD. This is a reconstruction of the Chinese mathematican's proof (based on his written instructions) that the sum of the squares on the sides of a right triangle equals the square on the hypotenuse. One begins with a2 and b2, the squares on the sides of the right triangle, and then cuts them into various shapes that can be rearranged to form c2, the square on the hypotenuse.

The Pythagorean theorem has fascinated people for nearly 4,000 years; there are now an estimated 367 different proofs, including ones by the Greek mathematician Pappus of Alexandria, the Arab mathematician-physician Thābit ibn Qurrah, the Italian artist-inventor Leonardo da Vinci, and even U.S. President James Garfield.

Trigonometric Functions

In mathematics, the trigonometric functions (also called circular functions, angle functions or goniometric functions) are real functions which relate an angle of a right-angled triangle to ratios of two side lengths. They are widely used in all sciences that are related to geometry, such as navigation, solid mechanics, celestial mechanics, geodesy, and many others. They are among the simplest periodic functions, and as such are also widely used for studying periodic phenomena, through Fourier analysis.

The most familiar trigonometric functions are the sine, the cosine, and the tangent. Their reciprocals are respectively the cosecant, the secant, and the cotangent, which are less used in modern mathematics.

The oldest definitions of trigonometric functions, related to right-angle triangles, define them only for acute angles. For extending these definitions to functions whose domain is the whole projectively extended real line, one can use geometrical definitions using the standard unit circle (a circle with radius 1 unit). Modern definitions express trigonometric functions as infinite series or as solutions of differential equations. This allows extending the domain of the sine and the cosine functions to the whole complex plane, and the domain of the other trigonometric functions to the complex plane from which some isolated points are removed.

Right-angled Triangle Definitions

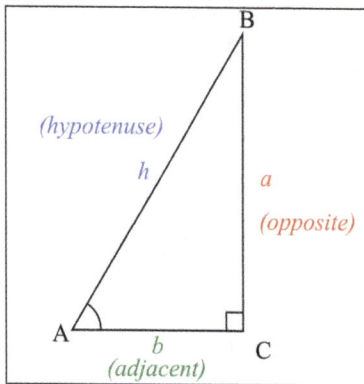

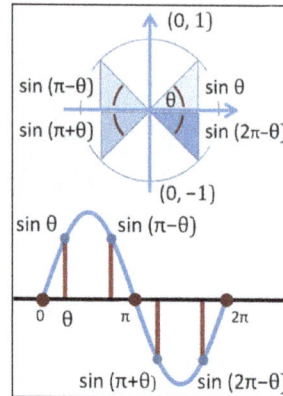

Top: Trigonometric function sin θ for selected angles θ, π − θ, π + θ, and 2π − θ in the four quadrants. Bottom: Graph of sine function versus angle. Angles from the top panel are identified.

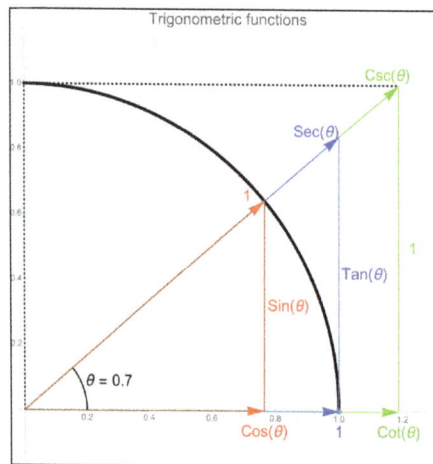

Plot of the six trigonometric functions and the unit circle for an angle of 0.7 radians.

In this topic, the same upper-case letter denotes a vertex of a triangle and the measure of the corresponding angle; the same lower case letter denotes an edge of the triangle and its length.

Given an acute angle A of a right-angled triangle the hypotenuse h is the side that connects the two acute angles. The side b *adjacent* to A is the side of the triangle that connects A to the right angle. The third side a is said *opposite* to A.

If the angle A is given, then all sides of the right-angled triangle are well defined up to a scaling factor. This means that the ratio of any two side lengths depends only on A. These six ratios define thus six functions of A, which are the trigonometric functions. More precisely, the six trigonometric functions are:

- Sine: $\sin A = \dfrac{a}{h} = \dfrac{\text{opposite}}{\text{hypotenuse}}$

- Cosine: $\cos A = \dfrac{b}{h} = \dfrac{\text{adjacent}}{\text{hypotenuse}}$

- Tangent: $\tan A = \dfrac{a}{b} = \dfrac{\text{opposite}}{\text{adjacent}}$

- Cosecant: $\csc A = \dfrac{h}{a} = \dfrac{\text{hypotenuse}}{\text{opposite}}$

- Secant: $\sec A = \dfrac{h}{b} = \dfrac{\text{hypotenuse}}{\text{adjacent}}$

- Cotangent: $\cot A = \dfrac{b}{a} = \dfrac{\text{adjacent}}{\text{opposite}}$

In a right angled triangle, the sum of the two acute angles is a right angle, that is 90° or $\dfrac{\pi}{2}$ radians.

This induces relationships between trigonometric functions that are summarized in the following table, where the angle is denoted by θ instead of A.

Function	Abbreviation	Description	Relationship (using radians)
Sine	sin	Opposite/ hypotenuse	$\sin\theta = \cos\left(\dfrac{\pi}{2}-\theta\right) = \dfrac{1}{\csc\theta}$
Cosine	cos	Adjacent/ hypotenuse	$\cos\theta = \sin\left(\dfrac{\pi}{2}-\theta\right) = \dfrac{1}{\sec\theta}$
Tangent	tan (or tg)	Opposite/adjacent	$\tan\theta = \dfrac{\sin\theta}{\cos\theta} = \cot\left(\dfrac{\pi}{2}-\theta\right) = \dfrac{1}{\cot\theta}$
Cotangent	cot (or cotan or cotg or ctg or ctn)	Adjacent/opposite	$\cot\theta = \dfrac{\cos\theta}{\sin\theta} = \tan\left(\dfrac{\pi}{2}-\theta\right) = \dfrac{1}{\tan\theta}$
Secant	sec	Hypotenuse/ adjacent	$\sec\theta = \csc\left(\dfrac{\pi}{2}-\theta\right) = \dfrac{1}{\cos\theta}$
Cosecant	csc (or cosec)	Hypotenuse/ opposite	$\csc\theta = \sec\left(\dfrac{\pi}{2}-\theta\right) = \dfrac{1}{\sin\theta}$

Radians versus Degrees

In geometric applications, the argument of a trigonometric function is generally the measure of an angle. For this purpose, any angular unit is convenient, and angles are most commonly measured in degrees.

When using trigonometric function in calculus, their argument is generally not an angle, but rather a real number. In this case, it is more suitable to express the argument of the trigonometric as the length of the arc of the unit circle delimited by an angle with the center of the circle as vertex. Therefore, one uses the radian as angular unit: a radian is the angle that delimits an arc of length 1 on the unit circle. A complete turn is thus an angle of 2π radians.

A great advantage of radians is that many formulas are much simpler when using them, typically all formulas relative to derivatives and integrals.

This is thus a general convention that, when the angular unit is not explicitly specified, the arguments of trigonometric functions are always expressed in radians.

Unit-circle Definitions

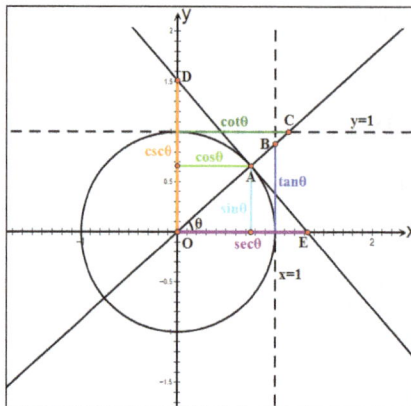

In this illustration, the six trigonometric functions of an arbitrary angle θ are represented as Cartesian coordinates of points related to the unit circle. The ordinates of A, B and D are $\sin\theta$, $\tan\theta$ and $\csc\theta$, respectively, while the abscissas of A, C and E are $\cos\theta$, $\cot\theta$ and $\sec\theta$, respectively.

The six trigonometric functions can be defined as coordinate values of points on the Euclidean plane that are related to the unit circle, which is the circle of radius one centered at the origin O of this coordinate system. While right-angled triangle definitions permit the definition of the trigonometric functions for angles between 0 and $\dfrac{\pi}{2}$ radian (90°), the unit circle definitions allow to extend the domain of the trigonometric functions to all positive and negative real numbers.

Quadrant II		Quadrant I	
"Science"		"All"	
sin, cosec	+	sin, cosec	+
cos, sec	−	cos, sec	+
tan, cot	−	tan, cot	+
Quadrant III		Quadrant IV	
"Teachers"		"Crazy"	
sin, cosec	−	sin, cosec	−
cos, sec	−	cos, sec	+
tan, cot	+	tan, cot	−

Signs of trigonometric functions in each quadrant. The mnemonic "all science teachers (are) crazy" lists the functions which are positive from quadrants I to IV. This is a variation on the mnemonic "All Students Take Calculus".

Rotating a ray from the direction of the positive half of the x-axis by an angle θ (counterclockwise for $\theta > 0$, and clockwise for $\theta < 0$) yields intersection points of this ray with the unit circle: $A = (x_A, y_A)$, and, by extending the ray to a line if necessary, with the line "$x = 1$": $B = (x_B, y_B)$, and with the line "$y = 1$": $C = (x_C, y_C)$. The tangent line to the unit circle in point A, which is orthogonal to this ray, intersects the y- and x-axis in points $D = (0, y_D)$ and $E = (x_E, 0)$. The coordinate values of these points give all the existing values of the trigonometric functions for arbitrary real values of θ in the following manner.

The trigonometric functions cos and sin are defined, respectively, as the x- and y-coordinate values of point A, i.e.,

$$\cos(\theta) = x_A \text{ and } \sin(\theta) = y_A$$

In the range $0 \le \theta \le \pi/2$ this definition coincides with the right-angled triangle definition by taking the right-angled triangle to have the unit radius OA as hypotenuse, and since for all points $P = (x, y)$ on the unit circle the equation $x^2 + y^2 = 1$ holds, this definition of cosine and sine also satisfies the Pythagorean identity:

$$\cos^2 \theta + \sin^2 \theta = 1$$

The other trigonometric functions can be found along the unit circle as:

$$\tan(\theta) = y_B \text{ and } \text{ot}(\theta) = x_C,$$

$$\cot(\theta) = y_D \text{ and } \sec(\theta) = x_E$$

By applying the Pythagorean identity and geometric proof methods, these definitions can readily be shown to coincide with the definitions of tangent, cotangent, secant and cosecant in terms of sine and cosine, that is:

$$\tan \theta = \frac{\sin \theta}{\cos \theta}, \quad \cot \theta = \frac{\cos \theta}{\sin \theta}, \quad \sec \theta = \frac{1}{\cos \theta}, \quad \csc \theta = \frac{1}{\sin \theta}.$$

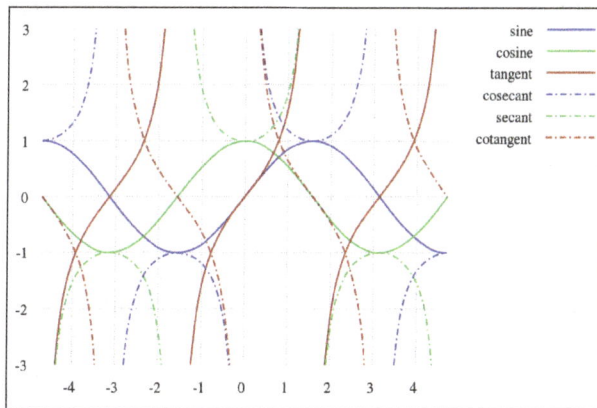

Trigonometric functions: Sine, Cosine, Tangent, Cosecant (dotted), Secant (dotted), Cotangent (dotted).

As a rotation of an angle of $\pm 2\pi$ does not change the position or size of a shape, the points A, B, C, D, and E are the same for two angles whose difference is an integer multiple of 2π. Thus trigonometric functions are periodic functions with period 2π. That is, the equalities,

$$\sin \theta = \sin(\theta + 2k\pi) \text{ and } \cos \theta = \cos(\theta + 2k\pi)$$

hold for any angle θ and any integer k. The same is true for the four other trigonometric functions. Observing the sign and the monotonicity of the functions sine, cosine, cosecant, and secant in the four quadrants, shows that 2π is the smallest value for which they are periodic, i.e., 2π is the fundamental period of these functions. However, already after a rotation by an angle θ the points B and C return to their original position, so that the tangent function and the cotangent function have a fundamental period of π. That is, the equalities:

$$\tan \theta = \tan(\theta + \pi) \text{ and } \cot \theta = \cot(\theta + k\pi)$$

hold for any angle θ and any integer k.

Algebraic Values

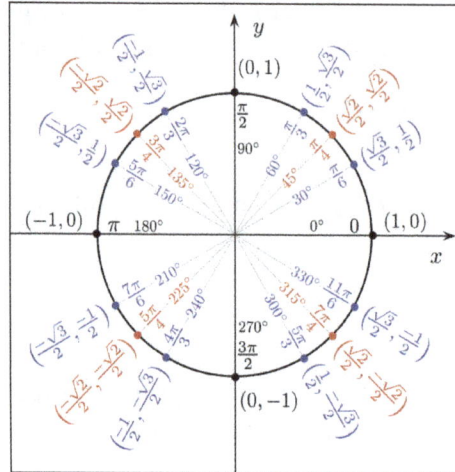

The unit circle, with some points labeled with their cosine and sine (in this order),
and the corresponding angles in radians and degrees.

The algebraic expressions for $\sin 0$, $\sin\dfrac{\pi}{6} = \sin 30°$, $\sin\dfrac{\pi}{4} = \sin 45°$, $\sin\dfrac{\pi}{3} = \sin 60°$ and $\sin\dfrac{\pi}{2} = \sin 90°$ are

$$0, \quad \frac{1}{2}, \quad \frac{\sqrt{2}}{2}, \quad \frac{\sqrt{3}}{2}, \quad 1,$$

respectively. Writing the numerators as square roots of consecutive natural numbers $\dfrac{\sqrt{0}}{2}, \dfrac{\sqrt{1}}{2}, \dfrac{\sqrt{2}}{2}, \dfrac{\sqrt{3}}{2}, \dfrac{\sqrt{4}}{2}$ provides an easy way to remember the values.

Such simple expressions generally do not exist for other angles which are rational multiples of a straight angle. For an angle which, measured in degrees, is a multiple of three, the sine and the cosine may be expressed in terms of square roots. These values of the sine and the cosine may thus be constructed by ruler and compass.

For an angle of an integer number of degrees, the sine and the cosine may be expressed in terms of square roots and the cube root of a non-real complex number. Galois theory allows proving that, if the angle is not a multiple of 3°, non-real cube roots are unavoidable.

For an angle which, measured in degrees, is a rational number, the sine and the cosine are algebraic numbers, which may be expressed in terms of nth roots. This results from the fact that the Galois groups of the cyclotomic polynomials are cyclic.

For an angle which, measured in degrees, is not a rational number, then either the angle or both the sine and the cosine are transcendental numbers. This is a corollary of Baker's theorem, proved in 1966.

Simple Algebraic Values

The following table summarizes the simplest algebraic values of trigonometric functions. The symbol ∞ represents the point at infinity on the projectively extended real line; it is not signed, because, when it appears in the table, the corresponding trigonometric function tends to $+\infty$ on one side, and to $-\infty$ on the other side, when the argument tends to the value in the table.

Radian	0	$\dfrac{\sqrt{6}-\sqrt{2}}{4}$	$\dfrac{\sqrt{2-\sqrt{2}}}{2}$	$\dfrac{1}{2}$	$\dfrac{\sqrt{2}}{2}$	$\dfrac{\sqrt{3}}{2}$	$\dfrac{\sqrt{6}+\sqrt{2}}{4}$	1
Degree	0°	15°	22.5°	30°	45°	60°	75°	90°
Sin	1	$\dfrac{\sqrt{6}+\sqrt{2}}{4}$	$\dfrac{\sqrt{2+\sqrt{2}}}{2}$	$\dfrac{\sqrt{3}}{2}$	$\dfrac{\sqrt{2}}{2}$	$\dfrac{1}{2}$	$\dfrac{\sqrt{6}-\sqrt{2}}{4}$	0
Cos	1	$\dfrac{\sqrt{6}+\sqrt{2}}{4}$	$\dfrac{\sqrt{2+\sqrt{2}}}{2}$	$\dfrac{\sqrt{3}}{2}$	$\dfrac{\sqrt{2}}{2}$	$\dfrac{1}{2}$	$\dfrac{\sqrt{6}-\sqrt{2}}{4}$	0
Tan	0	$2-\sqrt{3}$	$\sqrt{2}-1$	$\dfrac{\sqrt{3}}{3}$	1	$\sqrt{3}$	$2+\sqrt{3}$	∞
Cot	∞	$2+\sqrt{3}$	$\sqrt{2}+1$	$\sqrt{3}$	1	$\dfrac{\sqrt{3}}{3}$	$2-\sqrt{3}$	0
Sec	1	$\sqrt{6}-\sqrt{2}$	$\sqrt{2}\sqrt{2-\sqrt{2}}$	$\dfrac{2\sqrt{3}}{3}$	$\sqrt{2}$	2	$\sqrt{6}+\sqrt{2}$	∞
Csc	∞	$\sqrt{6}+\sqrt{2}$	$\sqrt{2}\sqrt{2+\sqrt{2}}$	2	$\sqrt{2}$	$\dfrac{2\sqrt{3}}{3}$	$\sqrt{6}-\sqrt{2}$	1

In Calculus

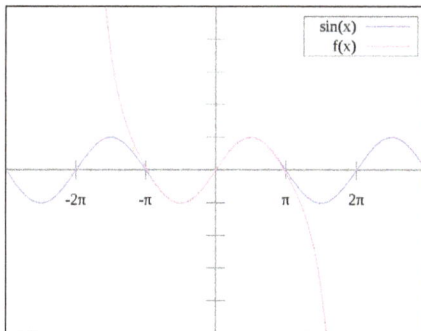

The sine function (blue) is closely approximated by its Taylor polynomial of degree 7 (pink) for a full cycle centered on the origin.

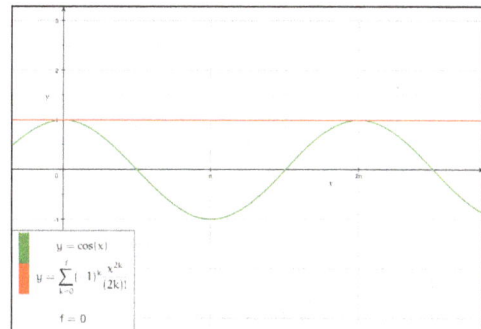

Animation for the approximation of cosine via Taylor polynomials.

Trigonometric functions are differentiable. This is not immediately evident from the above geometrical definitions. Moreover, the modern trend in mathematics is to build geometry from calculus rather than the converse. Therefore, except at a very elementary level, trigonometric functions are defined using the methods of calculus. For defining trigonometric functions inside calculus, there are two equivalent possibilities, either using power series or differential equations. These definitions are equivalent, as starting from one of them, it is easy to retrieve the other as a property. However the definition through differential equations is somehow more natural, since, for example, the choice of the coefficients of the power series may appear as quite arbitrary, and the Pythagorean identity is much easier to deduce from the differential equations.

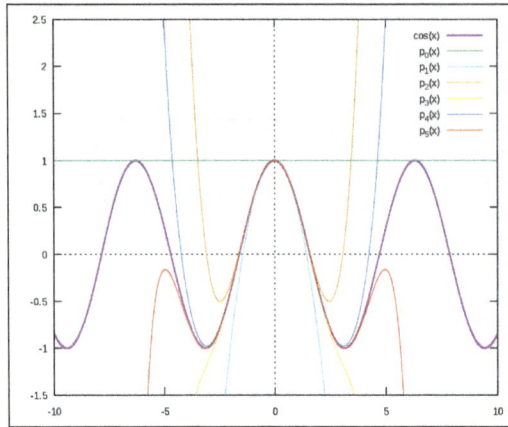

Together with the first Taylor polynomials.

Definition by Differential Equations

Sine and cosine are the unique differentiable functions such that:

$$\frac{d}{dx}\sin x = \cos x,$$

$$\frac{d}{dx}\cos x = -\sin x,$$

$$\sin 0 = 0,$$

$$\cos 0 = 1.$$

Differentiating these equations, one gets that both sine and cosine are solutions of the differential equation:

$$y'' + y = 0.$$

Applying the quotient rule to the definition of the tangent as the quotient of the sine by the cosine, one gets that the tangent function verifies:

$$\frac{d}{dx}\tan x = 1 + \tan^2 x.$$

Power Series Expansion

Applying the differential equations to power series with indeterminate coefficients, one may deduce recurrence relations for the coefficients of the Taylor series of the sine and cosine functions. These recurrence relations are easy to solve, and give the series expansions:

$$\sin x = x - \frac{x^3}{3!} + \frac{x^5}{5!} - \frac{x^7}{7!} +$$

$$\sum^{\infty} \frac{(\ 1)^n \quad {}^{2n \ 1}}{(2 \quad 1)!}$$

$$\cos x = 1 - \frac{x^2}{2!} + \frac{x^4}{4!} - \frac{x^6}{6!} + \cdots$$

$$= \sum_{n=0}^{\infty} \frac{(-1)^n x^{2n}}{(2n)!}$$

The radius of convergence of these series is infinite. Therefore, the sine and the cosine can be extended to entire functions (also called "sine" and "cosine"), which are complex-valued functions that are defined and holomorphic on the whole complex plane.

Being defined as fractions of entire functions, the other trigonometric functions may be extended to meromorphic functions, that is functions that are holomorphic in the whole complex plane, except some isolated points called poles. Here, the poles are the numbers of the form $(2k+1)\frac{\pi}{2}$ for the tangent and the secant, or $k\pi$ for the cotangent and the cosecant, where k is an arbitrary integer.

Recurrences relations may also be computed for the coefficients of the Taylor series of the other trigonometric functions. These series have a finite radius of convergence. Their coefficients have a combinatorial interpretation: they enumerate alternating permutations of finite sets.

More precisely, defining:

- U_n, the nth up/down number,

- B_n, the nth Bernoulli number,

- E_n, is the nth Euler number.

One has the following series expansions:

$$\tan x = \sum_{n=0}^{\infty} \frac{U_{2n+1} x^{2n+1}}{(2n+1)!}$$

$$= \sum_{n=1}^{\infty} \frac{(-1)^{n-1} 2^{2n} \left(2^{2n} - 1\right) B_{2n} x^{2n-1}}{(2n)!}$$

$$= x + \frac{1}{3} x^3 + \frac{2}{15} x^5 + \frac{17}{315} x^7 + \cdots, \qquad \text{for } |x| < \frac{\pi}{2}.$$

$$\csc x = \sum_{n=0}^{\infty} \frac{(-1)^{n+1} 2 \left(2^{2n-1} - 1\right) B_{2n} x^{2n-1}}{(2n)!}$$

$$= x^{-1} + \frac{1}{6} x + \frac{7}{360} x^3 + \frac{31}{15120} x^5 + \cdots, \qquad \text{for } 0 < |x| < \pi.$$

$$\sec x = \sum_{n=0}^{\infty} \frac{U_{2n} x^{2n}}{(2n)!} = \sum_{n=0}^{\infty} \frac{(-1)^n E_{2n} x^{2n}}{(2n)!}$$

$$= 1 + \frac{1}{2} x^2 + \frac{5}{24} x^4 + \frac{61}{720} x^6 + \cdots, \qquad \text{for } |x| < \frac{\pi}{2}.$$

$$\cot x = \sum_{n=0}^{\infty} \frac{(-1)^n 2^{2n} B_{2n} x^{2n-1}}{(2n)!}$$

$$= x^{-1} - \frac{1}{3}x - \frac{1}{45}x^3 - \frac{2}{945}x^5 - \cdots, \qquad \text{for } 0 < |x| < \pi.$$

There is a series representation as partial fraction expansion where just translated reciprocal functions are summed up, such that the poles of the cotangent function and the reciprocal functions match:

$$\pi \cdot \cot \pi x = \lim_{N \to \infty} \sum_{n=-N}^{N} \frac{1}{x+n}.$$

This identity can be proven with the Herglotz trick. Combining the $(-n)$th with the nth term lead to absolutely convergent series:

$$\pi \cdot \cot \pi x = \frac{1}{x} + 2x \sum_{n=1}^{\infty} \frac{1}{x^2 - n^2}, \qquad \frac{\pi}{\sin(\pi x)} = \frac{1}{x} + 2x \sum_{n=1}^{\infty} \frac{(-1)^n}{x^2 - n^2}.$$

Relationship to Exponential Function (Euler's Formula)

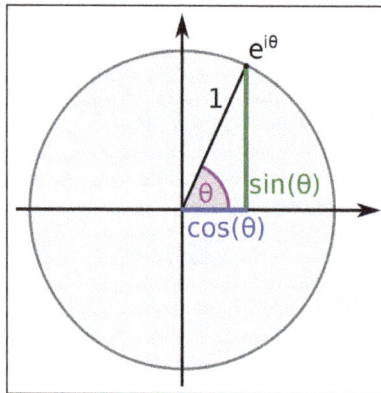

$\cos(\theta)$ and $\sin(\theta)$ are the real and imaginary part of $e^{i\theta}$ respectively.

Euler's formula relates sine and cosine to the exponential function:

$$e^{ix} = \cos x + i\sin x.$$

This formula is commonly considered for real values of x, but it remains true for all complex values.

Proof: Let $f_1(x) = \cos x + i \sin x$, and $f_2(x) = e^{ix}$. One has $\frac{d}{dx} f_j(x) = i f_j(x)$ for $j = 1, 2$. The quotient rule implies thus that $\frac{d}{dx}\left(\frac{f_1(x)}{f_2(x)}\right) = 0$. Therefore, $\frac{f_1(x)}{f_2(x)}$ is a constant function, which equals 1, as $f_1(0) = f_2(0) = 1$. This proves the formula.

One has:

$$e^{ix} = \cos x + i\sin x$$
$$e^{-ix} = \cos x - i\sin x.$$

Solving this linear system in sine and cosine, one can express them in terms of the exponential function:

$$\sin x = \frac{e^{ix} - e^{-ix}}{2i}$$

$$\cos x = \frac{e^{ix} + e^{-ix}}{2}.$$

When x is real, this may be rewritten as:

$$\cos x = \operatorname{Re}\left(e^{ix}\right), \qquad \sin x = \operatorname{Im}\left(e^{ix}\right).$$

Most trigonometric identities can be proved by expressing trigonometric functions in terms of the complex exponential function by using above formulas, and then using the identity $e^{a+b} = e^a e^b$ for simplifying the result.

Definitions using Functional Equations

One can also define the trigonometric functions using various functional equations.

For example, the sine and the cosine form the unique pair of continuous functions that satisfy the difference formula:

$$\cos(x - y) = \cos x \cos y + \sin x \sin y$$

and the added condition:

$$0 < x \cos x < \sin x < x \quad \text{for} \quad 0 < x < 1.$$

Basic Identities

Many identities interrelate the trigonometric functions. These identities may be proved geometrically from the unit-circle definitions or the right-angled-triangle definitions (although, for the latter definitions, care must be taken for angles that are not in the interval $[0, \pi/2]$. For non-geometrical proofs using only tools of calculus, one may use directly the differential equations, in a way that is similar to that of the above proof of Euler's identity. One can also use Euler's identity for expressing all trigonometric functions in terms of complex exponentials and using properties of the exponential function.

Parity

The cosine and the secant are even functions; the other trigonometric functions are odd functions. That is:

$$\sin(-x) = -\sin x$$
$$\cos(-x) = \cos x$$
$$\tan(-x) = -\tan x$$
$$\cot(-x) = -\cot x$$
$$\csc(-x) = -\csc x$$
$$\sec(-x) = \sec x.$$

Periods

All trigonometric functions are periodic functions of period 2π. This is the smallest period, except for the tangent and the cotangent, which have π as smallest period. This means that, for every integer k, one has:

$$\sin(x+2k\pi)=\sin x$$
$$\cos(x+2k\pi)=\cos x$$
$$\tan(x+k\pi)=\tan x$$
$$\cot(x+k\pi)=\cot x$$
$$\csc(x+2k\pi)=\csc x$$
$$\sec(x+2k\pi)=\sec x.$$

Pythagorean Identity

The Pythagorean identity, is the expression of the Pythagorean theorem in terms of trigonometric functions. It is:

$$\sin^2 x+\cos^2 x=1.$$

Sum and Difference Formulas

The sum and difference formulas allow expanding the sine, the cosine, and the tangent of a sum or a difference of two angles in terms of sines and cosines and tangents of the angles themselves. These can be derived geometrically, using arguments that date to Ptolemy. One can also produce them algebraically using Euler's formula.

Sum:

$$\sin\left(x+y\right)=\sin x\cos y+\cos x\sin y,$$
$$\cos\left(x+y\right)=\cos x\cos y-\sin x\sin y,$$
$$\tan(x+y)=\frac{\tan x+\tan y}{1-\tan x\tan y}.$$

Difference:

$$\sin\left(x-y\right)=\sin x\cos y-\cos x\sin y,$$
$$\cos\left(x-y\right)=\cos x\cos y+\sin x\sin y,$$
$$\tan(x-y)=\frac{\tan x-\tan y}{1+\tan x\tan y}.$$

When the two angles are equal, the sum formulas reduce to simpler equations known as the double-angle formulae.

$$\sin 2x = 2\sin x \cos x = \frac{2\tan x}{1+\tan^2 x},$$

$$\cos 2x = \cos^2 x - \sin^2 x = 2\cos^2 x - 1 = 1 - 2\sin^2 x = \frac{1-\tan^2 x}{1+\tan^2 x},$$

$$\tan 2x = \frac{2\tan x}{1-\tan^2 x}.$$

These identities can be used to derive the product-to-sum identities.

By setting $\theta = 2x$ and $t = \tan x$, this allows expressing all trigonometric functions of θ as a rational fraction of $t = \tan(\theta/2)$:

$$\sin\theta = \frac{2t}{1+t^2},$$

$$\cos\theta = \frac{1-t^2}{1+t^2},$$

$$\tan\theta = \frac{2t}{1-t^2}.$$

Together with:

$$d\theta = \frac{2}{1+t^2}\,dt,$$

This is the tangent half-angle substitution, which allows reducing the computation of integrals and antiderivatives of trigonometric functions to that of rational fractions.

Derivatives and Antiderivatives

The derivatives of trigonometric functions result from those of sine and cosine by applying quotient rule. The values given for the antiderivatives in the following formulas can be verified by differentiating them. The number C is a constant of integration.

$f(x)$	$f'(x)$	$\int f(x)dx$
$\sin x$	$\cos x$	$-\cos x + C$
$\cos x$	$-\sin x$	$\sin x + C$
$\tan x$	$\sec^2 x = 1 + \tan^2 x$	$-\ln(|\cos x|) + C$
$\cot x$	$-\csc^2 x = -(1 + \cot^2 x)$	$\ln(|\sin x|) + C$
$\sec x$	$\sec x \tan x$	$\ln(|\sec x + \tan x|) + C$
$\csc x$	$-\csc x \cot x$	$-\ln(|\csc x + \cot x|) + C$

Inverse Functions

The trigonometric functions are periodic, and hence not injective, so strictly speaking, they do not have an inverse function. However, on each interval on which a trigonometric function is monotonic, one can define an inverse function, and this defines inverse trigonometric functions as multivalued functions. To define a true inverse function, one must restrict the domain to an interval where the function is monotonic, and is thus bijective from this interval to its image by the function. The common choice for this interval, called the set of principal values, is given in the following table. As usual, the inverse trigonometric functions are denoted with the prefix "arc" before the name or its abbreviation of the function.

Function	Definition	Domain	Set of principal values
$y = \arcsin x$	$\sin y = x$	$-1 \le x \le 1$	$-\dfrac{\pi}{2} \le y \le \dfrac{\pi}{2}$
$y = \arccos x$	$\cos y = x$	$-1 \le x \le 1$	$0 \le y \le \pi$
$y = \arctan x$	$\tan y = x$	$-\infty \le x \le \infty$	$-\dfrac{\pi}{2} < y < \dfrac{\pi}{2}$
$y = \operatorname{arccot} x$	$\cot y = x$	$-\infty \le x \le \infty$	$0 < y < \pi$
$y = \operatorname{arcsec} x$	$\sec y = x$	$x < -1$ or $x > 1$	$0 \le y \le \pi,\, y \ne \dfrac{\pi}{2}$
$y = \operatorname{arccsc} x$	$\csc y = x$	$x < -1$ or $x > 1$	$-\dfrac{\pi}{2} \le y \le \dfrac{\pi}{2},\, y \ne 0$

The notations $\sin^{-1}$, $\cos^{-1}$ etc. are often used for arcsin and arccos, etc. When this notation is used, inverse functions could be confused with multiplicative inverses. The notation with the "arc" prefix avoids such a confusion, though "arcsec" for arcsecant can be confused with "arcsecond".

Just like the sine and cosine, the inverse trigonometric functions can also be expressed in terms of infinite series. They can also be expressed in terms of complex logarithms.

Applications

Angles and Sides of a Triangle

In this sections A, B, C denote the three (interior) angles of a triangle, and a, b, c denote the lengths of the respective opposite edges. They are related by various formulas, which are named by the trigonometric functions they involve.

Law of Sines

The law of sines states that for an arbitrary triangle with sides a, b, and c and angles opposite those sides A, B and C:

$$\frac{\sin A}{a} = \frac{\sin B}{b} = \frac{\sin C}{c} = \frac{2\Delta}{abc},$$

where Δ is the area of the triangle, or, equivalently,

$$\frac{a}{\sin A} = \frac{b}{\sin B} = \frac{c}{\sin C} = 2R,$$

where R is the triangle's circumradius.

It can be proven by dividing the triangle into two right ones and using the above definition of sine. The law of sines is useful for computing the lengths of the unknown sides in a triangle if two angles and one side are known. This is a common situation occurring in *triangulation*, a technique to determine unknown distances by measuring two angles and an accessible enclosed distance.

Law of Cosines

The law of cosines (also known as the cosine formula or cosine rule) is an extension of the Pythagorean theorem:

$$c^2 = a^2 + b^2 - 2ab\cos C,$$

or equivalently,

$$\cos C = \frac{a^2 + b^2 - c^2}{2ab}.$$

In this formula the angle at C is opposite to the side c. This theorem can be proven by dividing the triangle into two right ones and using the Pythagorean theorem.

The law of cosines can be used to determine a side of a triangle if two sides and the angle between them are known. It can also be used to find the cosines of an angle (and consequently the angles themselves) if the lengths of all the sides are known.

Law of Tangents

The following all form the law of tangents:

$$\frac{\tan\frac{A-B}{2}}{\tan\frac{A+B}{2}} = \frac{a-b}{a+b}; \qquad \frac{\tan\frac{A-C}{2}}{\tan\frac{A+C}{2}} = \frac{a-c}{a+c}; \qquad \frac{\tan\frac{B-C}{2}}{\tan\frac{B+C}{2}} = \frac{b-c}{b+c}$$

The explanation of the formulae in words would be cumbersome, but the patterns of sums and differences, for the lengths and corresponding opposite angles, are apparent in the theorem.

Law of Cotangents

If,

$$\zeta = \sqrt{\frac{1}{s}(s-a)(s-b)(s-c)} \quad \text{(the radius of the inscribed circle for the triangle)}$$

and

$$s = \frac{a+b+c}{2} \quad \text{(the semi-perimeter for the triangle),}$$

then the following all form the law of cotangents:

$$\cot\frac{A}{2} = \frac{s-a}{\zeta}; \qquad \cot\frac{B}{2} = \frac{s-b}{\zeta}; \qquad \cot\frac{C}{2} = \frac{s-c}{\zeta}$$

It follows that,

$$\frac{\cot\dfrac{A}{2}}{s-a} = \frac{\cot\dfrac{B}{2}}{s-b} = \frac{\cot\dfrac{C}{2}}{s-c}.$$

In words the theorem is: the cotangent of a half-angle equals the ratio of the semi-perimeter minus the opposite side to the said angle, to the inradius for the triangle.

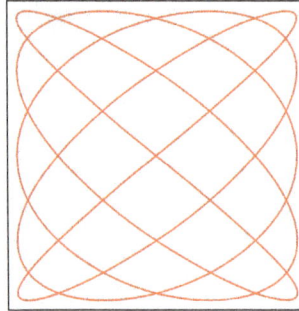

A Lissajous curve, a figure formed with a trigonometry-based function.

Periodic Functions

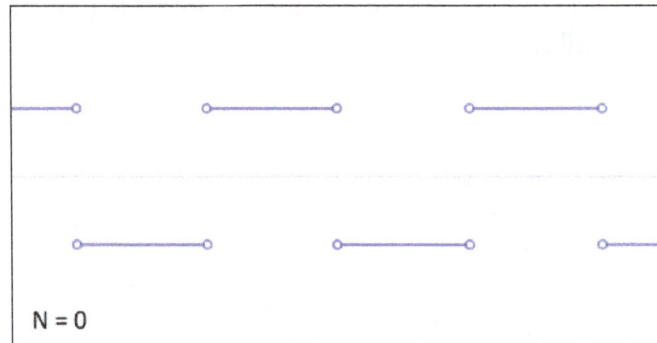

An animation of the additive synthesis of a square wave with an increasing number of harmonics.

The trigonometric functions are also important in physics. The sine and the cosine functions, for example, are used to describe simple harmonic motion, which models many natural phenomena, such as the movement of a mass attached to a spring and, for small angles, the pendular motion of a mass hanging by a string. The sine and cosine functions are one-dimensional projections of uniform circular motion.

Trigonometric functions also prove to be useful in the study of general periodic functions. The characteristic wave patterns of periodic functions are useful for modeling recurring phenomena such as sound or light waves.

Under rather general conditions, a periodic function $f(x)$ can be expressed as a sum of sine waves or cosine waves in a Fourier series. Denoting the sine or cosine basis functions by φ_k, the expansion of the periodic function $f(t)$ takes the form:

$$f(t) = \sum_{k=1}^{\infty} c_k \varphi_k(t).$$

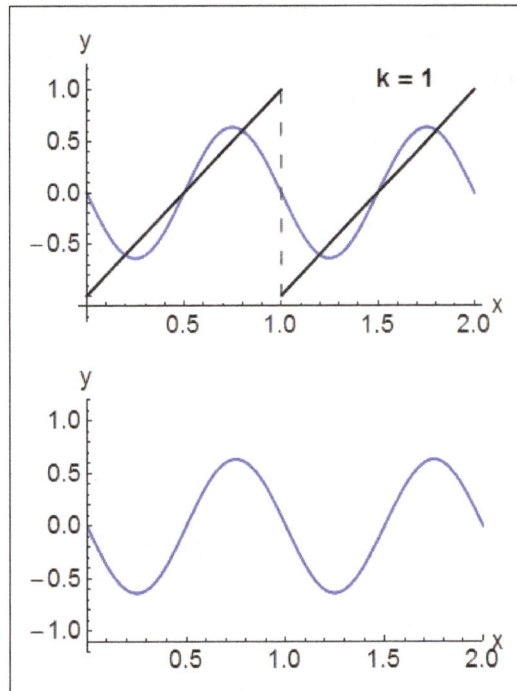

Sinusoidal basis functions (bottom) can form a sawtooth wave (top) when added. All the basis functions have nodes at the nodes of the sawtooth, and all but the fundamental ($k = 1$) have additional nodes. The oscillation seen about the sawtooth when k is large is called the Gibbs phenomenon.

For example, the square wave can be written as the Fourier series:

$$f_{\text{square}}(t) = \frac{4}{\pi}\sum_{k=1}^{\infty}\frac{\sin\big((2k-1)t\big)}{2k-1}.$$

In the animation of a square wave at top right it can be seen that just a few terms already produce a fairly good approximation. The superposition of several terms in the expansion of a sawtooth wave are shown underneath.

Inverse Trigonometric Functions

The inverse trigonometric relations are not functions because for any given input there exists more than one output. That is, for a given number there exists more than one angle whose sine, cosine, etc., is that number. The ranges of the inverse relations, however, can be restricted such that there is a one-to-one correspondence between the inputs and outputs of the inverse relations. With these restricted ranges, the inverse trigonometric relations become the inverse trigonometric functions.

The symbols for the inverse functions differ from the symbols for the inverse relations: the names of the functions are capitalized. The inverse functions appear as follows: Arcsine, Arccosine, Arctangent, Arccosecant, Arcsecant, and Arccotangent. They can also be represented like this: y = sin-1(x), y = cos-1(x), etc. The chart below shows the restricted ranges that transform the inverse relations into the inverse functions.

The domains of the inverse functions	
Inverse Trigonometric Function	Range
y = Arcsin (x)	$-\pi/2 \le y \le \pi/2$
y = Arccos (x)	$0 \le y \le \pi$
y = Arctan (x)	$-\pi/2 < y < \pi/2$
y = Arccsc (x)	$-\pi/2 \le y \le \pi/2 \ and \ y \ne 0$
y = Arcsec (x)	$0 \le y \le \pi \ and \ y \ne \pi/2$
y = Arccot (x)	$0 < y < \pi$

The inverse trigonometric functions do the same thing as the inverse trigonometric relations, but when an inverse functions is used, because of its restricted range, it only gives one output per input--whichever angle lies within its range. This creates a one-to-one correspondence and makes the inverse functions more usable and useful.

Knowledge of trigonometric and inverse trigonometric functions brings great power (and great responsibility). With knowledge of the trigonometric functions, we can calculate the value of a function at a given angle. With the inverse trigonometric functions, we can now calculate angles given certain function values. Solving both ways will be especially helpful as we attempt to solve triangles in the upcoming sections.

Applications of Trigonometry

Architecture and Engineering

Much of architecture and engineering relies on triangular supports. When an engineer determines the length of cables, the height of support towers, and the angle between the two when gauging weight loads and bridge strength, trigonometry helps him to calculate the correct angles. It also allows builders to correctly lay out a curved wall, figure the proper slope of a roof or the correct height and rise of a stairway. You can also use trigonometry at home to determine the height of a tree on your property without the need to climb dozens of feet in the air, or find the square footage of a curved piece of land.

Music Theory and Production

Trigonometry plays a major role in musical theory and production. Sound waves travel in a repeating wave pattern, which can be represented graphically by sine and cosine functions. A single note can be modeled on a sine curve, and a chord can be modeled with multiple sine curves used in conjunction with one another. A graphical representation of music allows computers to create and understand sounds. It also allows sound engineers to visualize sound waves so that they can adjust volume, pitch and other elements to create the desired sound effects. Trigonometry plays an important role in speaker placement as well, since the angles of sound waves hitting the ears can influence the sound quality.

Electrical Engineers and Trigonometry

Modern power companies use alternating current to send electricity over long-distance wires. In an alternating current, the electrical charge regularly reverses direction to deliver power safely and reliably to homes and businesses. Electrical engineers use trigonometry to model this flow and the change of direction, with the sine function used to model voltage. Every time you flip on a light switch or turn on the television, you're benefiting from one of trigonometry's many uses.

Manufacturing Industry

Trigonometry plays a major role in industry, where it allows manufacturers to create everything from automobiles to zigzag scissors. Engineers rely on trigonometric relationships to determine the sizes and angles of mechanical parts used in machinery, tools and equipment. This math plays a major role in automotive engineering, allowing car companies to size each part correctly and ensure they work safely together. Trigonometry is also used by seamstresses where determining the angle of darts or length of fabric needed to craft a certain shape of skirt or shirt is accomplished using basic trigonometric relationships.

References

- Bityutskov, V.I. (2011-02-07). "Trigonometric Functions". Encyclopedia of Mathematics. Archived from the original on 2017-12-29. Retrieved 2017-12-29

- Trigonometry, science: britannica.com, Retrieved 21 June, 2019

- Baragar, Arthur (2001), A Survey of Classical and Modern Geometries: With Computer Activities, Prentice Hall, Exercise 15.3, p. 301, ISBN 9780130143181

- Pythagorean-theorem, science, britannica.com, Retrieved 22 July, 2019

- Mitchell, Douglas W. (July 2001), "An Alternative Characterisation of All Primitive Pythagorean Triples", The Mathematical Gazette, 85 (503): 273–5, doi:10.2307/3622017, JSTOR 3622017

- Trigonometricequations, trigonometry, math: sparknotes.com, Retrieved 23 August, 2019

- O'Connor, J. J.; Robertson, E. F. "Madhava of Sangamagrama". School of Mathematics and Statistics University of St Andrews, Scotland. Archived from the original on 2006-05-14. Retrieved 2007-09-08

- Real-life-applications-trigonometry: sciencing.com, Retrieved 24 January, 2019

Probability and Statistics 6

- **Theories of Probability and Statistics**
- **Probability Axioms**

Probability helps to determine how likely an event can occur. The practice of collection, arrangement, presentation and analysis of numerical data is called statistics. This chapter delves into probability theory, statistical theory, decision theory, estimation theory, Bayes' theorem, probability axioms, etc. to provide an easy understanding of the subject.

Probability and statistics are the branches of mathematics concerned with the laws governing random events, including the collection, analysis, interpretation, and display of numerical data. Probability has its origin in the study of gambling and insurance in the 17th century, and it is now an indispensable tool of both social and natural sciences. Statistics may be said to have its origin in census counts taken thousands of years ago; as a distinct scientific discipline, however, it was developed in the early 19th century as the study of populations, economies, and moral actions and later in that century as the mathematical tool for analyzing such numbers.

Early Probability

Games of Chance

The modern mathematics of chance is usually dated to a correspondence between the French mathematicians Pierre de Fermat and Blaise Pascal in 1654. Their inspiration came from a problem about games of chance, proposed by a remarkably philosophical gambler, the chevalier de Méré. De Méré inquired about the proper division of the stakes when a game of chance is interrupted. Suppose two players, A and B, are playing a three-point game, each having wagered 32 pistoles, and are interrupted after A has two points and B has one. How much should each receive?

Fermat and Pascal proposed somewhat different solutions, though they agreed about the numerical answer. Each undertook to define a set of equal or symmetrical cases, then to answer the problem by comparing the number for A with that for B. Fermat, however, gave his answer in terms of the chances, or probabilities. He reasoned that two more games would suffice in any case to determine a victory. There are four possible outcomes, each equally likely in a fair game of chance. A might win twice, AA; or first A then B might win; or B then A; or BB. Of these four sequences, only the last would result in a victory for B. Thus, the odds for A are 3:1, implying a distribution of 48 pistoles for A and 16 pistoles for B.

Pascal thought Fermat's solution unwieldy, and he proposed to solve the problem not in terms of chances but in terms of the quantity now called "expectation." Suppose B had already won the next round. In that case, the positions of A and B would be equal, each having won two games, and each would be entitled to 32 pistoles. A should receive his portion in any case. B's 32, by contrast, depend on the assumption that he had won the first round. This first round can now be treated as a fair game for this stake of 32 pistoles, so that each player has an expectation of 16. Hence A's lot is 32 + 16, or 48, and B's is just 16.

Games of chance such as this one provided model problems for the theory of chances during its early period, and indeed they remain staples of the textbooks. A posthumous work of 1665 by Pascal on the "arithmetic triangle" now linked to his name binomial theorem showed how to calculate numbers of combinations and how to group them to solve elementary gambling problems. Fermat and Pascal were not the first to give mathematical solutions to problems such as these. More than a century earlier, the Italian mathematician, physician, and gambler Girolamo Cardano calculated odds for games of luck by counting up equally probable cases. His little book, however, was not published until 1663, by which time the elements of the theory of chances were already well known to mathematicians in Europe. It will never be known what would have happened had Cardano published in the 1520s. It cannot be assumed that probability theory would have taken off in the 16th century. When it began to flourish, it did so in the context of the "new science" of the 17th-century scientific revolution, when the use of calculation to solve tricky problems had gained a new credibility. Cardano, moreover, had no great faith in his own calculations of gambling odds, since he believed also in luck, particularly in his own. In the Renaissance world of monstrosities, marvels, and similitudes, chance—allied to fate—was not readily naturalized, and sober calculation had its limits.

Risks, Expectations and Fair Contracts

In the 17th century, Pascal's strategy for solving problems of chance became the standard one. It was, for example, used by the Dutch mathematician Christiaan Huygens in his short treatise on games of chance, published in 1657. Huygens refused to define equality of chances as a fundamental presumption of a fair game but derived it instead from what he saw as a more basic notion of an equal exchange. Most questions of probability in the 17th century were solved, as Pascal solved his, by redefining the problem in terms of a series of games in which all players have equal expectations. The new theory of chances was not, in fact, simply about gambling but also about the legal notion of a fair contract. A fair contract implied equality of expectations, which served as the fundamental notion in these calculations. Measures of chance or probability were derived secondarily from these expectations.

Probability was tied up with questions of law and exchange in one other crucial respect. Chance and risk, in aleatory contracts, provided a justification for lending at interest, and hence a way of avoiding Christian prohibitions against usury. Lenders, the argument went, were like investors; having shared the risk, they deserved also to share in the gain. For this reason, ideas of chance had already been incorporated in a loose, largely nonmathematical way into theories of banking and marine insurance. From about 1670, initially in the Netherlands, probability began to be used to determine the proper rates at which to sell annuities. Jan de Wit, leader of the Netherlands from 1653 to 1672, corresponded in the 1660s with Huygens, and eventually he published a small treatise on the subject of annuities in 1671.

Annuities in early modern Europe were often issued by states to raise money, especially in times of war. They were generally sold according to a simple formula such as "seven years purchase," meaning that the annual payment to the annuitant, promised until the time of his or her death, would be one-seventh of the principal. This formula took no account of age at the time the annuity was purchased. Wit lacked data on mortality rates at different ages, but he understood that the proper charge for an annuity depended on the number of years that the purchaser could be expected to live and on the presumed rate of interest. Despite his efforts and those of other mathematicians, it remained rare even in the 18th century for rulers to pay much heed to such quantitative considerations. Life insurance, too, was connected only loosely to probability calculations and mortality records, though statistical data on death became increasingly available in the course of the 18th century. The first insurance society to price its policies on the basis of probability calculations was the Equitable, founded in London in 1762.

Probability as the Logic of Uncertainty

The English clergyman Joseph Butler, in his very influential Analogy of Religion, called probability "the very guide of life." The phrase, however, did not refer to mathematical calculation but merely to the judgments made where rational demonstration is impossible. The word probability was used in relation to the mathematics of chance in 1662 in the Logic of Port-Royal, written by Pascal's fellow Jansenists, Antoine Arnauld and Pierre Nicole. But from medieval times to the 18th century and even into the 19th, a probable belief was most often merely one that seemed plausible, came on good authority, or was worthy of approval. Probability, in this sense, was emphasized in England and France from the late 17th century as an answer to skepticism. Man may not be able to attain perfect knowledge but can know enough to make decisions about the problems of daily life. The new experimental natural philosophy of the later 17th century was associated with this more modest ambition, one that did not insist on logical proof.

Almost from the beginning, however, the new mathematics of chance was invoked to suggest that decisions could after all be made more rigorous. Pascal invoked it in the most famous chapter of his Pensées, "Of the Necessity of the Wager," in relation to the most important decision of all, whether to accept the Christian faith. One cannot know of God's existence with absolute certainty; there is no alternative but to bet. Perhaps, he supposed, the unbeliever can be persuaded by consideration of self-interest. If there is a God (Pascal assumed he must be the Christian God), then to believe in him offers the prospect of an infinite reward for infinite time. However small the probability, provided only that it be finite, the mathematical expectation of this wager is infinite. For so great a benefit, one sacrifices rather little, perhaps a few paltry pleasures during one's brief life on Earth. It seemed plain which was the more reasonable choice.

The link between the doctrine of chance and religion remained an important one through much of the 18th century, especially in Britain. Another argument for belief in God relied on a probabilistic natural theology. The classic instance is a paper read by John Arbuthnot to the Royal Society of London in 1710 and published in its Philosophical Transactions in 1712. Arbuthnot presented there a table of christenings in London from 1629 to 1710. He observed that in every year there was a slight excess of male over female births. The proportion, approximately 14 boys for every 13 girls, was perfectly calculated, given the greater dangers to which young men are exposed in their search for food, to bring the sexes to an equality of numbers at the age of marriage. Could this excellent result have been produced by chance alone? Arbuthnot thought not, and he deployed a probability calculation to demonstrate the point. The probability that male births would by accident exceed female ones in 82 consecutive years is $(0.5)82$. Considering further that this excess is found all over the world, he said, and within fixed limits of variation, the chance becomes almost infinitely small. This argument for the overwhelming probability of Divine Providence was repeated by many—and refined by a few. The Dutch natural philosopher Willem's Gravesande incorporated the limits of variation of these birth ratios into his mathematics and so attained a still more decisive vindication of Providence over chance. Nicolas Bernoulli, from the famous Swiss mathematical family, gave a more skeptical view. If the underlying probability of a male birth was assumed to be 0.5169 rather than 0.5, the data were quite in accord with probability theory. That is, no Providential direction was required.

Apart from natural theology, probability came to be seen during the 18th-century Enlightenment as a mathematical version of sound reasoning. In 1677 the German mathematician Gottfried Wilhelm Leibniz imagined a utopian world in which disagreements would be met by this challenge: "Let us calculate, Sir." The French mathematician Pierre-Simon de Laplace, in the early 19th century, called probability "good sense reduced to calculation." This ambition, bold enough, was not quite so scientific as it may first appear. For there were some cases where a straightforward application of probability mathematics led to results that seemed to defy rationality. One example, proposed by Nicolas Bernoulli and made famous as the St. Petersburg paradox, involved a bet with an exponentially increasing payoff. A fair coin is to be tossed until the first time it comes up heads. If it comes up heads on the first toss, the payment is 2 ducats; if the first time it comes up heads is on the second toss, 4 ducats; and if on the nth toss, 2n ducats. The mathematical expectation of this game is infinite, but no sensible person would pay a very large sum for the privilege of receiving the payoff from it. The disaccord between calculation and reasonableness created a problem, addressed by generations of mathematicians. Prominent among them was Nicolas's cousin Daniel Bernoulli, whose solution depended on the idea that a ducat added to the wealth of a rich man benefits him much less than it does a poor man.

Probability arguments figured also in more practical discussions, such as debates during the 1750s and '60s about the rationality of smallpox inoculation. Smallpox was at this time widespread and deadly, infecting most and carrying off perhaps one in seven Europeans. Inoculation in these days involved the actual transmission of smallpox, not the cowpox vaccines developed in the 1790s by the English surgeon Edward Jenner, and was itself moderately risky. Was it rational to accept a small probability of an almost immediate death to reduce greatly a large probability of death by smallpox in the indefinite future? Calculations of mathematical expectation, as by Daniel Bernoulli, led unambiguously to a favourable answer. But some disagreed, most famously the eminent mathematician and perpetual thorn in the flesh of probability theorists, the French mathematician Jean Le Rond d'Alembert. One might, he argued, reasonably prefer a greater assurance of surviving in the near term to improved prospects late in life.

The Probability of Causes

Many 18th-century ambitions for probability theory, including Arbuthnot's, involved reasoning from effects to causes. Jakob Bernoulli, uncle of Nicolas and Daniel, formulated and proved a law of large numbers to give formal structure to such reasoning. This was published in 1713 from a manuscript, the Ars conjectandi, left behind at his death in 1705. There he showed that the ob-served proportion of, say, tosses of heads or of male births will converge as the number of trials increases to the true probability p, supposing that it is uniform. His theorem was designed to give assurance that when p is not known in advance, it can properly be inferred by someone with suffi-cient experience. He thought of disease and the weather as in some way like drawings from an urn. At bottom they are deterministic, but since one cannot know the causes in sufficient detail, one must be content to investigate the probabilities of events under specified conditions.

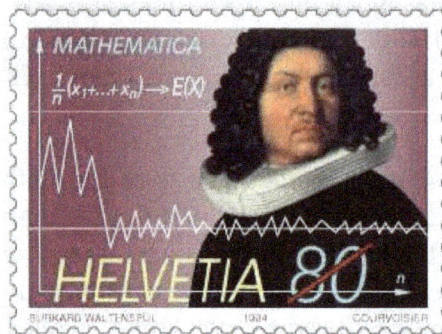

Swiss commemorative stamp of mathematician Jakob Bernoulli, issued 1994, displaying the formula and the graph for the law of large numbers, first proved by Bernoulli in 1713.

The English physician and philosopher David Hartley announced in his Observations on Man that a certain "ingenious Friend" had shown him a solution of the "inverse problem" of reasoning from the occurrence of an event p times and its failure q times to the "original Ratio" of causes. But Hartley named no names, and the first publication of the formula he promised occurred in 1763 in a posthumous paper of Thomas Bayes, communicated to the Royal Society by the British philoso-pher Richard Price. This has come to be known as Bayes's theorem. But it was the French, especial-ly Laplace, who put the theorem to work as a calculus of induction, and it appears that Laplace's publication of the same mathematical result in 1774 was entirely independent. The result was perhaps more consequential in theory than in practice. An exemplary application was Laplace's probability that the sun will come up tomorrow, based on 6,000 years or so of experience in which it has come up every day.

Laplace and his more politically engaged fellow mathematicians, most notably Marie-Jean-An-toine-Nicolas de Caritat, marquis de Condorcet, hoped to make probability into the foundation of the moral sciences. This took the form principally of judicial and electoral probabilities, address-ing thereby some of the central concerns of the Enlightenment philosophers and critics. Justice and elections were, for the French mathematicians, formally similar. In each, a crucial question was how to raise the probability that a jury or an electorate would decide correctly. One element involved testimonies, a classic topic of probability theory. In 1699 the British mathematician John Craig used probability to vindicate the truth of scripture and, more idiosyncratically, to forecast the end of time, when, due to the gradual attrition of truth through successive testimonies, the Christian religion would become no longer probable. The Scottish philosopher David Hume, more skeptically, argued in probabilistic but nonmathematical language beginning in 1748 that the tes-timonies supporting miracles were automatically suspect, deriving as they generally did from un-educated persons, lovers of the marvelous. Miracles, moreover, being violations of laws of nature, had such a low a priori probability that even excellent testimony could not make them probable. Condorcet also wrote on the probability of miracles, or at least faits extraordinaires, to the end of subduing the irrational. But he took a more sustained interest in testimonies at trials, proposing to weigh the credibility of the statements of any particular witness by considering the proportion of times that he had told the truth in the past, and then use inverse probabilities to combine the testimonies of several witnesses.

Laplace and Condorcet applied probability also to judgments. In contrast to English juries, French juries voted whether to convict or acquit without formal deliberations. The probabilists began by supposing that the jurors were independent and that each had a probability p greater than 1/2 of reaching a true verdict. There would be no injustice, Condorcet argued, in exposing innocent defendants to a risk of conviction equal to risks they voluntarily assume without fear, such as crossing the English Channel from Dover to Calais. Using this number and considering also the interest of the state in minimizing the number of guilty who go free, it was possible to calculate an optimal jury size and the majority required to convict. This tradition of judicial probabilities lasted into the 1830s, when Laplace's student Siméon-Denis Poisson used the new statistics of criminal justice to measure some of the parameters. But by this time the whole enterprise had come to seem gravely doubtful, in France and elsewhere. In 1843 the English philosopher John Stuart Mill called it "the opprobrium of mathematics," arguing that one should seek more reliable knowledge rather than waste time on calculations that merely rearrange ignorance.

The Rise of Statistics

Political Arithmetic

During the 19th century, statistics grew up as the empirical science of the state and gained preeminence as a form of social knowledge. Population and economic numbers had been collected, though often not in a systematic way, since ancient times and in many countries. In Europe the late 17th century was an important time also for quantitative studies of disease, population, and wealth. In 1662 the English statistician John Graunt published a celebrated collection of numbers and observations pertaining to mortality in London, using records that had been collected to chart the advance and decline of the plague. In the 1680s the English political economist and statistician William Petty published a series of essays on a new science of "political arithmetic," which combined statistical records with bold—some thought fanciful—calculations, such as, for example, of the monetary value of all those living in Ireland. These studies accelerated in the 18th century and were increasingly supported by state activity, though ancien régime governments often kept the numbers secret. Administrators and savants used the numbers to assess and enhance state power but also as part of an emerging "science of man." The most assiduous, and perhaps the most renowned, of these political arithmeticians was the Prussian pastor Johann Peter Süssmilch, whose study of the divine order in human births and deaths was first published in 1741 and grew to three fat volumes by 1765. The decisive proof of Divine Providence in these demographic affairs was their regularity and order, perfectly arranged to promote man's fulfillment of what he called God's first commandment, to be fruitful and multiply. Still, he did not leave such matters to nature and to God, but rather he offered abundant advice about how kings and princes could promote the growth of their populations. He envisioned a rather spartan order of small farmers, paying modest rents and taxes, living without luxury, and practicing the Protestant faith. Roman Catholicism was unacceptable on account of priestly celibacy.

Social Numbers

Lacking, as they did, complete counts of population, 18th-century practitioners of political arithmetic had to rely largely on conjectures and calculations. In France especially, mathematicians such as Laplace used probability to surmise the accuracy of population figures determined from samples. In the 19th century such methods of estimation fell into disuse, mainly because they were replaced by regular, systematic censuses. The census of the United States, required by the U.S. Constitution and conducted every 10 years beginning in 1790, was among the earliest.

Sweden had begun earlier; most of the leading nations of Europe followed by the mid-19th century. They were also eager to survey the populations of their colonial possessions, which indeed were among the very first places to be counted. A variety of motives can be identified, ranging from the requirements of representative government to the need to raise armies. Some of this counting can scarcely be attributed to any purpose, and indeed the contemporary rage for numbers was by no means limited to counts of human populations. From the mid-18th century and especially after the conclusion of the Napoleonic Wars in 1815, the collection and publication of numbers proliferated in many domains, including experimental physics, land surveys, agriculture, and studies of the weather, tides, and terrestrial magnetism. For perhaps the best statistical graph ever constructed. Still, the management of human populations played a decisive role in the statistical enthusiasm of the early 19th century. Political instabilities associated with the French Revolution of 1789 and the economic changes of early industrialization made social science a great desideratum. A new field of moral statistics grew up to record and comprehend the problems of dirt, disease, crime, ignorance, and poverty.

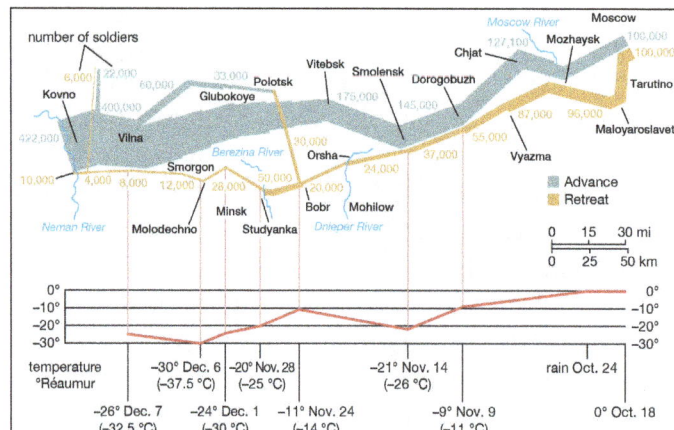

Statistical map of Napoleon's Russian campaign.

The size of Napoleon's army during the Russian campaign of 1812 is shown by the dwindling width of the lines of advance (green) and retreat (gold). The retreat information is correlated with a temperature scale shown along the lower portion of the statistical map.

Some of these investigations were conducted by public bureaus, but much was the work of civic-minded professionals, industrialists, and, especially after midcentury, women such as Florence Nightingale. One of the first serious statistical organizations arose in 1832 as section F of the new British Association for the Advancement of Science. The intellectual ties to natural science were uncertain at first, but there were some influential champions of statistics as a mathematical science. The most effective was the Belgian mathematician Adolphe Quetelet, who argued untiringly that mathematical probability was essential for social statistics. Quetelet hoped to create from these materials a new science, which he called at first social mechanics and later social physics. He wrote often of the analogies linking this science to the most mathematical of the natural sciences, celestial mechanics. In practice, though, his methods were more like those of geodesy or meteorology, involving massive collections of data and the effort to detect patterns that might be identified as laws. These, in fact, seemed to abound. He found them in almost every collection of social numbers, beginning with some publications of French criminal statistics from the mid-1820s. The numbers, he announced, were essentially constant from year to year, so steady that one could speak here of statistical laws. If there was something paradoxical in these "laws" of crime, it was nonetheless comforting to find regularities underlying the manifest disorder of social life.

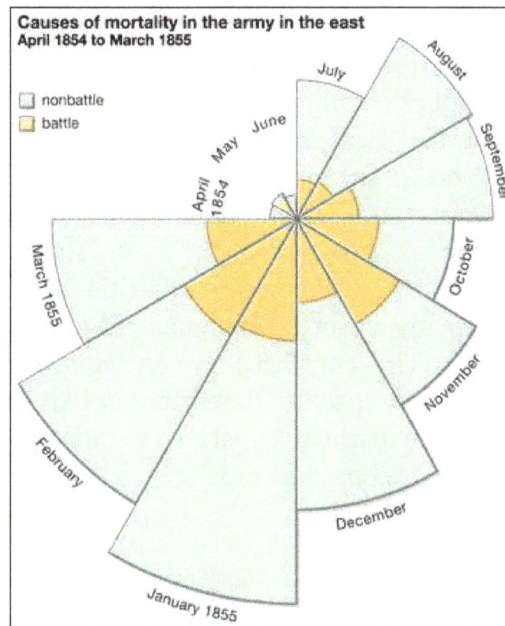

Coxcomb chart: The English nurse Florence Nightingale was an innovator in displaying statistical data through graphs. In 1858 she devised the type depicted here, which she named Coxcomb. Like pie charts, the Coxcomb indicates frequency by relative area, but it differs in its use of fixed angles and variable radii.

New Kind of Regularity

Even Quetelet had been startled at first by the discovery of these statistical laws. Regularities of births and deaths belonged to the natural order and so were unsurprising, but here was constancy of moral and immoral acts, acts that would normally be attributed to human free will. Was there some mysterious fatalism that drove individuals, even against their will, to fulfill a budget of crimes? Were such actions beyond the reach of human intervention? Quetelet determined that they were not. Nevertheless, he continued to emphasize that the frequencies of such deeds should be understood in terms of causes acting at the level of society, not of choices made by individuals. His view was challenged by moralists, who insisted on complete individual responsibility for thefts, murders, and suicides. Quetelet was not so radical as to deny the legitimacy of punishment, since the system of justice was thought to help regulate crime rates. Yet he spoke of the murderer on the scaffold as himself a victim, part of the sacrifice that society requires for its own conservation. Individually, to be sure, it was perhaps within the power of the criminal to resist the inducements that drove him to his vile act. Collectively, however, crime is but trivially affected by these individual decisions. Not criminals but crime rates form the proper object of social investigation.

Reducing them is to be achieved not at the level of the individual but at the level of the legislator, who can improve society by providing moral education or by improving systems of justice. Statisticians have a vital role as well. To them falls the task of studying the effects on society of legislative changes and of recommending measures that could bring about desired improvements.

Quetelet's arguments inspired a modest debate about the consistency of statistics with human free will. This intensified after 1857, when the English historian Henry Thomas Buckle recited his favourite examples of statistical law to support an uncompromising determinism in his immensely successful History of Civilization in England. Interestingly, probability had been linked to deterministic arguments from very early in its history, at least since the time of Jakob Bernoulli. Laplace argued in his Philosophical Essay on Probabilities that man's dependence on probability was simply a consequence of imperfect knowledge. A being who could follow every particle in the universe, and who had unbounded powers of calculation, would be able to know the past and to predict the future with perfect certainty. The statistical determinism inaugurated by Quetelet had a quite different character. Now it was not necessary to know things in infinite detail. At the microlevel, indeed, knowledge often fails, for who can penetrate the human soul so fully as to comprehend why a troubled individual has chosen to take his or her own life? Yet such uncertainty about individuals somehow dissolves in light of a whole society, whose regularities are often more perfect than those of physical systems such as the weather. Not real persons but l'homme moyen, the average man, formed the basis of social physics. This contrast between individual and collective phenomena was, in fact, hard to reconcile with an absolute determinism like Buckle's. Several critics of his book pointed this out, urging that the distinctive feature of statistical knowledge was precisely its neglect of individuals in favour of mass observations.

Statistical Physics

The same issues were discussed also in physics. Statistical understandings first gained an influential role in physics at just this time, in consequence of papers by the German mathematical physicist Rudolf Clausius from the late 1850s and, especially, of one by the Scottish physicist James Clerk Maxwell published in 1860. Maxwell, at least, was familiar with the social statistical tradition, and he had been sufficiently impressed by Buckle's History and by the English astronomer John Herschel's influential essay on Quetelet's work in the Edinburgh Review to discuss them in letters. During the1870s, Maxwell often introduced his gas theory using analogies from social statistics. The first point, a crucial one, was that statistical regularities of vast numbers of molecules were quite sufficient to derive thermodynamic laws relating the pressure, volume, and temperature in gases. Some physicists, including, for a time, the German Max Planck, were troubled by the contrast between a molecular chaos at the microlevel and the very precise laws indicated by physical instruments. They wondered if it made sense to seek a molecular, mechanical grounding for thermodynamic laws. Maxwell invoked the regularities of crime and suicide as analogies to the statistical laws of thermodynamics and as evidence that local uncertainty can give way to large-scale predictability. At the same time, he insisted that statistical physics implied a certain imperfection of knowledge. In physics, as in social science, determinism was very much an issue in the 1850s and '60s. Maxwell argued that physical determinism could only be speculative, since human knowledge of events at the molecular level is necessarily imperfect. Many of the laws of physics, he said, are like those regularities detected by census officers: they are quite sufficient as a guide to practical life, but they lack the certainty characteristic of abstract dynamics.

The Spread of Statistical Mathematics

Statisticians, wrote the English statistician Maurice Kendall in 1942, "have already overrun every branch of science with a rapidity of conquest rivaled only by Attila, Mohammed, and the Colorado beetle." The spread of statistical mathematics through the sciences began, in fact, at least a century before there were any professional statisticians. Even regardless of the use of probability to estimate populations and make insurance calculations, this history dates back at least to 1809. In that year, the German mathematician Carl Friedrich Gauss published a derivation of the new method of least squares incorporating a mathematical function that soon became known as the astronomer's curve of error, and later as the Gaussian or normal distribution.

The problem of combining many astronomical observations to give the best possible estimate of one or several parameters had been discussed in the 18th century. The first publication of the method of least squares as a solution to this problem was inspired by a more practical problem, the analysis of French geodetic measures undertaken in order to fix the standard length of the metre. This was the basic measure of length in the new metric system, decreed by the French Revolution and defined as 1/40,000,000 of the longitudinal circumference of the Earth. In 1805 the French mathematician Adrien-Marie Legendre proposed to solve this problem by choosing values that minimize the sums of the squares of deviations of the observations from a point, line, or curve drawn through them. In the simplest case, where all observations were measures of a single point, this method was equivalent to taking an arithmetic mean.

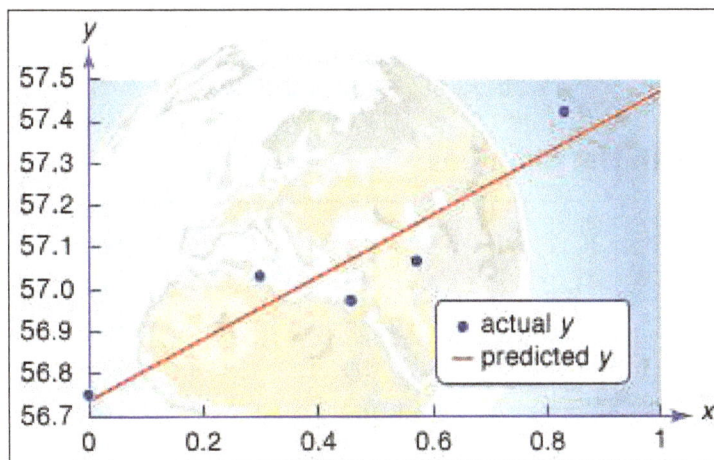

Measuring the shape of the Earth using the least squares approximationThe graph is based on measurements taken about 1750 near Rome by mathematician Ruggero Boscovich. The x-axis covers one degree of latitude, while the y-axis corresponds to the length of the arc along the meridian as measured in units of Paris toise (=1.949 metres). The straight line represents the least squares approximation, or average slope, for the measured data, allowing the mathematician to predict arc lengths at other latitudes and thereby calculate the shape of the Earth.

Gauss soon announced that he had already been using least squares since 1795, a somewhat doubtful claim. After Legendre's publication, Gauss became interested in the mathematics of least squares, and he showed in 1809 that the method gave the best possible estimate of a parameter if the errors of the measurements were assumed to follow the normal distribution. This distribution, whose importance for mathematical probability and statistics was decisive, was first shown by the

French mathematician Abraham de Moivre in the 1730s to be the limit (as the number of events increases) for the binomial distribution. In particular, this meant that a continuous function (the normal distribution) and the power of calculus could be substituted for a discrete function (the binomial distribution) and laborious numerical methods. Laplace used the normal distribution extensively as part of his strategy for applying probability to very large numbers of events. The most important problem of this kind in the 18th century involved estimating populations from smaller samples. Laplace also had an important role in reformulating the method of least squares as a problem of probabilities. For much of the 19th century, least squares was overwhelmingly the most important instance of statistics in its guise as a tool of estimation and the measurement of uncertainty. It had an important role in astronomy, geodesy, and related measurement disciplines, including even quantitative psychology. Later, about 1900, it provided a mathematical basis for a broader field of statistics that came to be used by a wide range of fields.

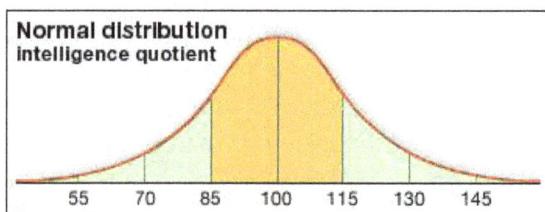

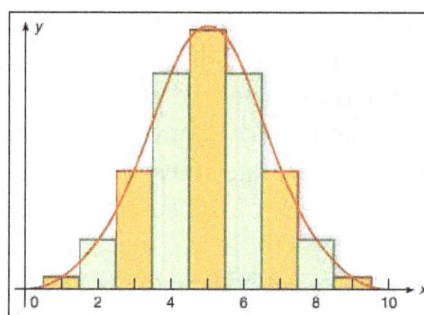

Grap of intelligence quotient (IQ) as a normal distribution with a mean of 100 and a standard deviation of 15. The shaded between 85 and 115 (within one standard deviation of the mean) accounts for about 68 percent of the total area, hence 68 percent of all IQ scores.

Statistical Theories in the Sciences

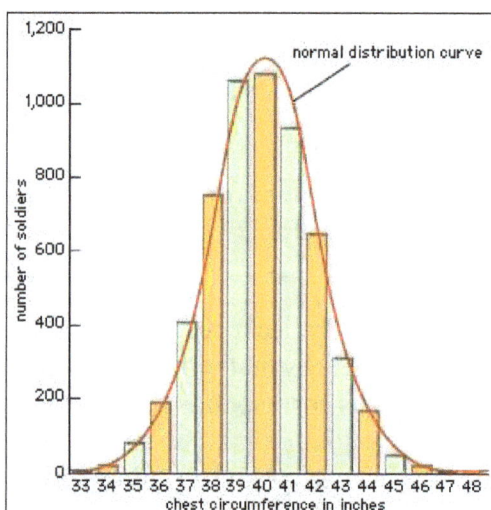

Histogram (bar chart) showing chest measurements of 5,732 Scottish soldiers This was the first time that a human characteristic had been shown to follow a normal distribution, as indicated by the superimposed curve.

The role of probability and statistics in the sciences was not limited to estimation and measurement. Equally significant, and no less important for the formation of the mathematical field, were statistical theories of collective phenomena that bypassed the study of individuals. The social science bearing the name statistics was the prototype of this approach. Quetelet advanced its mathematical level by incorporating the normal distribution into it. He argued that human traits of every sort, from chest circumference and height to the distribution of propensities to marry or commit crimes, conformed to the astronomer's error law. The kinetic theory of gases of Clausius, Maxwell, and the Austrian physicist Ludwig Boltzmann was also a statistical one. Here it was not the imprecision or uncertainty of scientific measurements but the motions of the molecules themselves to which statistical understandings and probabilistic mathematics were applied. Once again, the error law played a crucial role. The Maxwell-Boltzmann distribution law of molecular velocities, as it has come to be known, is a three-dimensional version of this same function. In importing it into physics, Maxwell drew both on astronomical error theory and on Quetelet's social physics.

Biometry

The English biometric school developed from the work of the polymath Francis Galton, cousin of Charles Darwin. Galton admired Quetelet, but he was critical of the statistician's obsession with mean values rather than variation. The normal law, as he began to call it, was for him a way to measure and analyze variability. This was especially important for studies of biological evolution, since Darwin's theory was about natural selection acting on natural diversity. A figure from Galton's 1877 paper on breeding sweet peas shows a physical model, now known as the Galton board, that he employed to explain the normal distribution of inherited characteristics; in particular, he used his model to explain the tendency of progeny to have the same variance as their parents, a process he called reversion, subsequently known as regression to the mean. Galton was also founder of the eugenics movement, which called for guiding the evolution of human populations the same way that breeders improve chickens or cows. He developed measures of the transmission of parental characteristics to their offspring: the children of exceptional parents were generally somewhat exceptional themselves, but there was always, on average, some reversion or regression toward the population mean. He developed the elementary mathematics of regression and correlation as a theory of hereditary transmission and thus as statistical biological theory rather than as a mathematical tool. However, Galton came to recognize that these methods could be applied to data in many fields, and by 1889, when he published his Natural Inheritance, he stressed the flexibility and adaptability of his statistical tools.

Still, evolution and eugenics remained central to the development of statistical mathematics. The most influential site for the development of statistics was the biometric laboratory set up at University College London by Galton's admirer, the applied mathematician Karl Pearson. From about 1892 he collaborated with the English biologist Walter F.R. Weldon on quantitative studies of evolution, and he soon began to attract an assortment of students from many countries and disciplines who hoped to learn the new statistical methods. Their journal, Biometrika, was for many years the most important venue for publishing new statistical tools and for displaying their uses.

Biometry was not the only source of new developments in statistics at the turn of the 19th century. German social statisticians such as Wilhelm Lexis had turned to more mathematical approaches some decades earlier. In England, the economist Francis Edgeworth became interested in statistical mathematics in the early 1880s. One of Pearson's earliest students, George Udny Yule, turned away from biometry and especially from eugenics in favour of the statistical investigation of so-cial data. Nevertheless, biometry provided an important model, and many statistical techniques, for other disciplines. The 20th-century fields of psychometrics, concerned especially with mental testing, and econometrics, which focused on economic time-series, reveal this relationship in their very names.

Samples and Experiments

Near the beginning of the 20th century, sampling regained its respectability in social statistics, for reasons that at first had little to do with mathematics. Early advocates, such as the first direc-tor of the Norwegian Central Bureau of Statistics, A.N. Kiaer, thought of their task primarily in terms of attaining representativeness in relation to the most important variables—for example, geographic region, urban and rural, rich and poor. The London statistician Arthur Bowley was among the first to urge that sampling should involve an element of randomness. Jerzy Ney-man, a statistician from Poland who had worked for a time in Pearson's laboratory, wrote a particularly decisive mathematical paper on the topic in 1934. His method of stratified sampling incorporated a concern for representativeness across the most important variables, but it also required that the individuals sampled should be chosen randomly. This was designed to avoid selection biases but also to create populations to which probability theory could be applied to calculate expected errors. George Gallup achieved fame in 1936 when his polls, employing strat-ified sampling, successfully predicted the reelection of Franklin Delano Roosevelt, in defiance of the Literary Digest's much larger but uncontrolled survey, which forecast a landslide for the Republican Alfred Landon.

The alliance of statistical tools and experimental design was also largely an achievement of the 20th century. Here, too, randomization came to be seen as central. The emerging protocol called for the establishment of experimental and control populations and for the use of chance where possible to decide which individuals would receive the experimental treatment. These exper-imental repertoires emerged gradually in educational psychology during the 1900s and '10s. They were codified and given a full mathematical basis in the next two decades by Ronald A. Fisher, the most influential of all the 20th-century statisticians. Through randomized, controlled experiments and statistical analysis, he argued, scientists could move beyond mere correlation to causal knowledge even in fields whose phenomena are highly complex and variable. His ideas of experimental design and analysis helped to reshape many disciplines, including psychology, ecology, and therapeutic research in medicine, especially during the triumphant era of quantifi-cation after 1945.

The Modern Role of Statistics

In some ways, statistics has finally achieved the Enlightenment aspiration to create a logic of un-certainty. Statistical tools are at work in almost every area of life, including agriculture, business, engineering, medicine, law, regulation, and social policy, as well as in the physical, biological, and social sciences and even in parts of the academic humanities. The replacement of human "com-puters" with mechanical and then electronic ones in the 20th century greatly lightened the im-mense burdens of calculation that statistical analysis once required. Statistical tests are used to as-sess whether observed results, such as increased harvests where fertilizer is applied, or improved earnings where early childhood education is provided, give reasonable assurance of causation, rather than merely random fluctuations. Following World War II, these significance levels virtually came to define an acceptable result in some of the sciences and also in policy applications.

From about 1930 there grew up in Britain and America—and a bit later in other countries—a profession of statisticians, experts in inference, who defined standards of experimentation as well as methods of analysis in many fields. To be sure, statistics in the various disciplines retained a fair degree of specificity. There were also divergent schools of statisticians, who disagreed, often vehemently, on some issues of fundamental importance. Fisher was highly critical of Pearson; Neyman and Egon Pearson, while unsympathetic to father Karl's methods, disagreed also with Fisher's. Under the banner of Bayesianism appeared yet another school, which, against its predecessors, emphasized the need for subjective assessments of prior probabilities. The most immoderate ambitions for statistics as the royal road to scientific inference depended on unacknowledged compromises that ignored or dismissed these disputes. Despite them, statistics has thrived as a somewhat heterogeneous but powerful set of tools, methods, and forms of expertise that continues to regulate the acquisition and interpretation of quantitative data.

Theories of Probability and Statistics

Probability Theory

Probability theory is a branch of mathematics concerned with the analysis of random phenomena. The outcome of a random event cannot be determined before it occurs, but it may be any one of several possible outcomes. The actual outcome is considered to be determined by chance.

The word probability has several meanings in ordinary conversation. Two of these are particularly important for the development and applications of the mathematical theory of probability. One is the interpretation of probabilities as relative frequencies, for which simple games involving coins, cards, dice, and roulette wheels provide examples. The distinctive feature of games of chance is that the outcome of a given trial cannot be predicted with certainty, although the collective results of a large number of trials display some regularity.

For example, the statement that the probability of "heads" in tossing a coin equals one-half, according to the relative frequency interpretation, implies that in a large number of tosses the relative frequency with which "heads" actually occurs will be approximately one-half, although it contains no implication concerning the outcome of any given toss. There are many similar examples involving groups of people, molecules of a gas, genes, and so on. Actuarial statements about the life expectancy for persons of a certain age describe the collective experience of a large number of individuals but do not purport to say what will happen to any particular person. Similarly, predictions about the chance of a genetic disease occurring in a child of parents having a known genetic makeup are statements about relative frequencies of occurrence in a large number of cases but are not predictions about a given individual.

Mathematical concepts of probability theory, illustrated by some of the applications that have stimulated their development.

Experiments, Sample Space, Events and Equally likely Probabilities

Applications of Simple Probability Experiments

The fundamental ingredient of probability theory is an experiment that can be repeated, at least hypothetically, under essentially identical conditions and that may lead to different outcomes on different trials. The set of all possible outcomes of an experiment is called a "sample space." The experiment of tossing a coin once results in a sample space with two possible outcomes, "heads" and "tails." Tossing two dice has a sample space with 36 possible outcomes, each of which can be identified with an ordered pair (i, j), where i and j assume one of the values 1, 2, 3, 4, 5, 6 and denote the faces showing on the individual dice. It is important to think of the dice as identifiable (say by a difference in colour), so that the outcome (1, 2) is different from (2, 1). An "event" is a well-defined subset of the sample space. For example, the event "the sum of the faces showing on the two dice equals six" consists of the five outcomes (1, 5), (2, 4), (3, 3), (4, 2), and (5, 1).

A third example is to draw n balls from an urn containing balls of various colours. A generic outcome to this experiment is an n-tuple, where the ith entry specifies the colour of the ball obtained on the ith draw (i = 1, 2,..., n). In spite of the simplicity of this experiment, a thorough understanding gives the theoretical basis for opinion polls and sample surveys.

For example, individuals in a population favouring a particular candidate in an election may be identified with balls of a particular colour, those favouring a different candidate may be identified with a different colour, and so on. Probability theory provides the basis for learning about the contents of the urn from the sample of balls drawn from the urn; an application is to learn about the electoral preferences of a population on the basis of a sample drawn from that population.

Another application of simple urn models is to use clinical trials designed to determine whether a new treatment for a disease, a new drug, or a new surgical procedure is better than a standard treatment. In the simple case in which treatment can be regarded as either success or failure, the goal of the clinical trial is to discover whether the new treatment more frequently leads to success than does the standard treatment. Patients with the disease can be identified with balls in an urn. The red balls are those patients who are cured by the new treatment, and the black balls are those not cured.

Usually there is a control group, who receive the standard treatment. They are represented by a second urn with a possibly different fraction of red balls. The goal of the experiment of drawing some number of balls from each urn is to discover on the basis of the sample which urn has the larger fraction of red balls. A variation of this idea can be used to test the efficacy of a new vaccine. Perhaps the largest and most famous example was the test of the Salk vaccine for poliomyelitis conducted in 1954. It was organized by the U.S. Public Health Service and involved almost two million children. Its success has led to the almost complete elimination of polio as a health problem in the industrialized parts of the world. Strictly speaking, these applications are problems of statistics, for which the foundations are provided by probability theory.

In contrast to the experiments described, many experiments have infinitely many possible outcomes. For example, one can toss a coin until "heads" appears for the first time. The number of possible tosses is n = 1, 2,.... Another example is to twirl a spinner. For an idealized spinner made from a straight line segment having no width and pivoted at its centre, the set of possible outcomes is the set of all angles that the final position of the spinner makes with some fixed direction, equivalently all real numbers in $[0, 2\pi)$. Many measurements in the natural and social sciences, such as volume, voltage, temperature, reaction time, marginal income, and so on, are made on continuous scales and at least in theory involve infinitely many possible values. If the repeated measurements on different subjects or at different times on the same subject can lead to different outcomes, probability theory is a possible tool to study this variability.

Now suppose that a coin is tossed n times, and consider the probability of the event "heads does not occur" in the n tosses. An outcome of the experiment is an n-tuple, the kth entry of which identifies the result of the kth toss. Since there are two possible outcomes for each toss, the number of elements in the sample space is 2^n. Of these, only one outcome corresponds to having no heads, so the required probability is $1/2^n$.

It is only slightly more difficult to determine the probability of "at most one head." In addition to the single case in which no head occurs, there are n cases in which exactly one head occurs, because it can occur on the first, second,..., or nth toss. Hence, there are n + 1 cases favourable to obtaining at most one head, and the desired probability is $(n + 1)/2^n$.

The Principle of Additivity

This last example illustrates the fundamental principle that, if the event whose probability is sought can be represented as the union of several other events that have no outcomes in common ("at most one head" is the union of "no heads" and "exactly one head"), then the probability of the union is the sum of the probabilities of the individual events making up the union. To describe this situation symbolically, let S denote the sample space. For two events A and B, the intersection of A and B is the set of all experimental outcomes belonging to both A and B and is denoted $A \cap B$; the union of A and B is the set of all experimental outcomes belonging to A or B (or both) and is denoted $A \cup B$. The impossible event—i.e., the event containing no outcomes—is denoted by $\varnothing$. The probability of an event A is written P(A). The principle of addition of probabilities is that, if A_1, A_2,..., A_n are events with $A_i \cap A_j = \varnothing$ for all pairs $i \neq j$, then:

$$P(A_1 \cup A_2 \cup \cdots \cup A_n) = P(A_1) + \ldots + P(A_n)$$

Equation above is consistent with the relative frequency interpretation of probabilities; for, if $A_i \cap A_j = \varnothing$ for all $i \neq j$, the relative frequency with which at least one of the A_i occurs equals the sum of the relative frequencies with which the individual A_i occur.

Equation above is fundamental for everything that follows. Indeed, in the modern axiomatic theory of probability, which eschews a definition of probability in terms of "equally likely outcomes" as being hopelessly circular, an extended form of equation above plays a basic role.

An elementary, useful consequence of equation above is the following. With each event A is associated the complementary event A^c consisting of those experimental outcomes that do not belong to A. Since $A \cap A^c = \varnothing$, $A \cup A^c = S$, and P(S) = 1 (where S denotes the sample space), it follows from equation above that $P(A^c) = 1 - P(A)$. For example, the probability of "at least one head" in n tosses of a coin is one minus the probability of "no head," or $1 - 1/2^n$.

Multinomial Probability

A basic problem first solved by Jakob Bernoulli is to find the probability of obtaining exactly i red balls in the experiment of drawing n times at random with replacement from an urn containing b black and r red balls. To draw at random means that, on a single draw, each of the r + b balls is equally likely to be drawn and, since each ball is replaced before the next draw, there are $(r + b) \times \cdots \times (r + b) = (r + b)^n$ possible outcomes to the experiment. Of these possible outcomes, the number that is favourable to obtaining i red balls and n − i black balls in any one particular order is,

$$\overbrace{r \times r \times \ldots \times r}^{i} \times \overbrace{b \times b \times \ldots \times b}^{n-i} = r^i \times b^{n-i}$$

The number of possible orders in which i red balls and n − i black balls can be drawn from the urn is the binomial coefficient,

$$\binom{n}{i} = \frac{n!}{i!(n-i)!},$$

where k! = k × (k − 1) ×⋯× 2 × 1 for positive integers k, and 0! = 1. Hence, the probability in question, which equals the number of favourable outcomes divided by the number of possible outcomes, is given by the binomial distribution,

$$\binom{n}{i} = \frac{r^i b^{n-i}}{(r+b)^n} = \binom{n}{i} p^i q^{n-i} \qquad (i = 0, 1, 2, \ldots, n),$$

where p = r/(r + b) and q = b/(r + b) = 1 − p.

For example, suppose r = 2b and n = 4. According to above equation, the probability of "exactly two red balls" is:

$$\binom{4}{2}\left(\frac{2}{3}\right)^2 \left(\frac{1}{3}\right)^2 = 6 \times \frac{4}{81} = \frac{8}{27}.$$

In this case the,

$$\binom{4}{2} = 6$$

possible outcomes are easily enumerated: (rrbb), (rbrb), (brrb), (rbbr), (brbr), (bbrr).

For a derivation of equation $\binom{n}{i} = \frac{n!}{i!(n-i)!}$, observe that in order to draw exactly i red balls in n draws one must either draw i red balls in the first n − 1 draws and a black ball on the nth draw or draw i − 1 red balls in the first n − 1 draws followed by the ith red ball on the nth draw. Hence,

$$\binom{n}{i} = \binom{n-1}{i} + \binom{n-1}{i-1},$$

from which equation $\binom{n}{i} = \frac{n!}{i!(n-i)!}$, can be verified by induction on n.

Two related examples are (i) drawing without replacement from an urn containing r red and b black balls and (ii) drawing with or without replacement from an urn containing balls of s different colours. If n balls are drawn without replacement from an urn containing r red and b black balls, the number of possible outcomes is,

$$\binom{r+b}{n},$$

of which the number favourable to drawing i red and n − i black balls is:

$$\binom{r}{i}\binom{b}{n-1}.$$

Hence, the probability of drawing exactly i red balls in n draws is the ratio,

$$\frac{\binom{r}{i}\binom{b}{n-1}}{\binom{r+b}{n}}$$

If an urn contains balls of s different colours in the ratios $p_1:p_2:...:ps$, where $p_1 +\cdots+ ps = 1$ and if n balls are drawn with replacement, the probability of obtaining i_1 balls of the first colour, i_2 balls of the second colour, and so on is the multinomial probability,

$$\frac{n!}{i_1!i_2!...i_s!}p_1^{i_1} p_2^{i_2}\cdots p_s^{i_s}$$

The evaluation of equation $\binom{n}{i}=\dfrac{r^i b^{n-i}}{(r+b)^n}=\binom{n}{i}p^i q^{n-i}$ $(i=0, 1, 2,\ldots, n)$, with pencil and paper grows increasingly difficult with increasing n. It is even more difficult to evaluate related cumulative probabilities—for example the probability of obtaining "at most j red balls" in the n draws, which can be expressed as the sum of equation for i = 0, 1,..., j. The problem of approximate computation of probabilities that are known in principle is a recurrent theme throughout the history of probability theory.

The Birthday Problem

An entertaining example is to determine the probability that in a randomly selected group of n people at least two have the same birthday. If one assumes for simplicity that a year contains 365 days and that each day is equally likely to be the birthday of a randomly selected person, then in a group of n people there are 365n possible combinations of birthdays. The simplest solution is to determine the probability of no matching birthdays and then subtract this probability from 1. Thus, for no matches, the first person may have any of the 365 days for his birthday, the second any of the remaining 364 days for his birthday, the third any of the remaining 363 days,..., and the nth any of the remaining 365 − n + 1. The number of ways that all n people can have different birthdays is then 365 × 364 ×···× (365 − n + 1), so that the probability that at least two have the same birthday is,

$$P=1-\frac{365\times364\times\ldots\times(365-n+1)}{365^n}.$$

Numerical evaluation shows, rather surprisingly, that for n = 23 the probability that at least two people have the same birthday is about 0.5 (half the time). For n = 42 the probability is about 0.9 (90 percent of the time).

This example illustrates that applications of probability theory to the physical world are facilitated by assumptions that are not strictly true, although they should be approximately true. Thus, the assumptions that a year has 365 days and that all days are equally likely to be the birthday of a random individual are false, because one year in four has 366 days and because birth dates are not distributed uniformly throughout the year. Moreover, if one attempts to apply this result to an actual group of individuals, it is necessary to ask what it means for these to be "randomly selected." It would naturally be unreasonable to apply it to a group known to contain twins. In spite of the obvious failure of the assumptions to be literally true, as a classroom example, it rarely disappoints instructors of classes having more than 40 students.

Conditional Probability

Suppose two balls are drawn sequentially without replacement from an urn containing r red and b black balls. The probability of getting a red ball on the first draw is r/(r + b). If, however, one is told that a red ball was obtained on the first draw, the conditional probability of getting a red ball on the second draw is (r − 1)/(r + b − 1), because for the second draw there are r + b − 1 balls in the urn, of which r − 1 are red. Similarly, if one is told that the first ball drawn is black, the conditional probability of getting red on the second draw is r/(r + b − 1).

In a number of trials the relative frequency with which B occurs among those trials in which A occurs is just the frequency of occurrence of A ∩ B divided by the frequency of occurrence of A. This suggests that the conditional probability of B given A (denoted P(B|A)) should be defined by:

$$P(B \backslash A) = \frac{P(A \cap B)}{P(A)}$$

If A denotes a red ball on the first draw and B a red ball on the second draw in the experiment of the preceding paragraph, then P(A) = r/(r + b) and:

$$P(A \cap B) = \frac{r(r-1)}{\left[(r+b)(r+b-1)\right]},$$

which is consistent with the "obvious" answer derived above.

Rewriting equation $P(B \backslash A) = \dfrac{P(A \cap B)}{P(A)}$ as P(A ∩ B) = P(A)P(B|A) and adding to this expression the same expression with A replaced by Ac ("not A") leads via $P(A_1 \cup A_2 \cup \cdots \cup A_n) = P(A_1) + \ldots + P(A_n)$ to the equality,

$$P(B) = P(A \cap B) + (A^c \cap B)$$

$$= P(A)P(B|A) + P(A^c)P(B|A^c).$$

More generally, if A_1, A_2,..., A_n are mutually exclusive events and their union is the entire sample space, so that exactly one of the A_k must occur, essentially the same argument gives a fundamental relation, which is frequently called the law of total probability:

$$P(B) = P(A_1)P(B|A_1) + P(A_2)P(B|A_2) + \cdots + P(A_n)P(B|A_n).$$

Applications of Conditional Probability

An application of the law of total probability to a problem originally posed by Christiaan Huygens is to find the probability of "gambler's ruin." Suppose two players, often called Peter and Paul, initially have x and m − x dollars, respectively. A ball, which is red with probability p and black with probability q = 1 − p, is drawn from an urn. If a red ball is drawn, Paul must pay Peter one dollar, while Peter must pay Paul one dollar if the ball drawn is black. The ball is replaced, and the game continues until one of the players is ruined. It is quite difficult to determine the probability of Peter's ruin by a direct analysis of all possible cases. But let Q(x) denote that probability as a function of Peter's initial fortune x and observe that after one draw the structure of the rest of the game is exactly as it was before the first draw, except that Peter's fortune is now either x + 1 or x − 1 according to the results of the first draw. The law of total probability with A = {red ball on first draw} and A^c = {black ball on first draw} shows that:

$$Q(x) = pQ(x+1) + qQ(x-1).$$

This equation holds for x = 2, 3,..., m − 2. It also holds for x = 1 and m − 1 if one adds the boundary conditions Q(0) = 1 and Q(m) = 0, which say that if Peter has 0 dollars initially, his probability of ruin is 1, while if he has all m dollars, he is certain to win.

It can be verified by direct substitution that above equation together with the indicated boundary conditions are satisfied by:

$$Q(x) = \frac{\left(\frac{q}{p}\right)^x - \left(\frac{q}{p}\right)^m}{1 - \left(\frac{q}{p}\right)^m} \left(p \neq \tfrac{1}{2}\right)$$

$$= 1 - \frac{x}{m}\left(p = \tfrac{1}{2}\right)$$

With some additional analysis it is possible to show that these give the only solutions and hence must be the desired probabilities.

Suppose m = 10x, so that Paul initially has nine times as much money as Peter. If p = 1/2, the probability of Peter's ruin is 0.9 regardless of the values of x and m. If p = 0.51, so that each trial slightly favours Peter, the situation is quite different. For x = 1 and m = 10, the probability of Peter's ruin is 0.88, only slightly less than before. However, for x = 100 and m = 1,000, Peter's slight advantage on each trial becomes so important that the probability of his ultimate ruin is now less than 0.02.

Generalizations of the problem of gambler's ruin play an important role in statistical sequential analysis, developed by the Hungarian-born American statistician Abraham Wald in response to the demand for more efficient methods of industrial quality control during World War II. They also enter into insurance risk theory.

The following example shows that, even when it is given that A occurs, it is important in evaluating P(B|A) to recognize that Ac might have occurred, and hence in principle it must be possible also to evaluate P(B|A^c). By lot, two out of three prisoners—Sam, Jean, and Chris—are chosen to be executed. There are:

$$\binom{3}{2} = 6$$

possible pairs of prisoners to be selected for execution, of which two contain Sam, so the probability that Sam is slated for execution is 2/3. Sam asks the guard which of the others is to be executed. Since at least one must be, it appears that the guard would give Sam no information by answering. After hearing that Jean is to be executed, Sam reasons that, since either he or Chris must be the other one, the conditional probability that he will be executed is 1/2. Thus, it appears that the guard has given Sam some information about his own fate. However, the experiment is incompletely defined, because it is not specified how the guard chooses whether to answer "Jean" or "Chris" in case both of them are to be executed. If the guard answers "Jean" with probability p, the conditional probability of the event "Sam will be executed" given "the guard says Jean will be executed" is:

$$\frac{\frac{1}{3}}{\frac{1}{3}+\frac{P}{3}} = \frac{1}{1+P}$$

Only in the case p = 1 is Sam's reasoning correct. If p = 1/2, the guard in fact gives no information about Sam's fate.

Independence

One of the most important concepts in probability theory is that of "independence." The events A and B are said to be (stochastically) independent if P(B|A) = P(B), or equivalently if

$$P(A \cap B) = P(A)P(B)$$

The intuitive meaning of the definition in terms of conditional probabilities is that the probability of B is not changed by knowing that A has occurred. Equation above shows that the definition is symmetric in A and B.

It is intuitively clear that, in drawing two balls with replacement from an urn containing r red and b black balls, the event "red ball on the first draw" and the event "red ball on the second draw" are independent. (This statement presupposes that the balls are thoroughly mixed before each draw.) An analysis of the $(r + b)^2$ equally likely outcomes of the experiment shows that the formal definition is indeed satisfied.

In terms of the concept of independence, the experiment leading to the binomial distribution can be described as follows. On a single trial a particular event has probability p. An experiment consists of n independent repetitions of this trial. The probability that the particular event occurs exactly i times is given by equation:

$$\binom{n}{i} = \frac{r^i b^{n-i}}{(r+b)^n} = \binom{n}{i} p^i q^{n-i} \qquad (i = 0, 1, 2, \ldots, n)$$

Independence plays a central role in the law of large numbers, the central limit theorem, the Poisson distribution, and Brownian motion.

Bayes's Theorem

Consider now the defining relation for the conditional probability $P(A_n|B)$, where the Ai are mutually exclusive and their union is the entire sample space. Substitution of $P(A_n)P(B|A_n)$ in the numerator of equation $P(B\backslash A) = \dfrac{P(A \cap B)}{P(A)}$ and substitution of the right-hand side of the law of total probability in the denominator yields a result known as Bayes's theorem (after the 18th-century English clergyman Thomas Bayes) or the law of inverse probability:

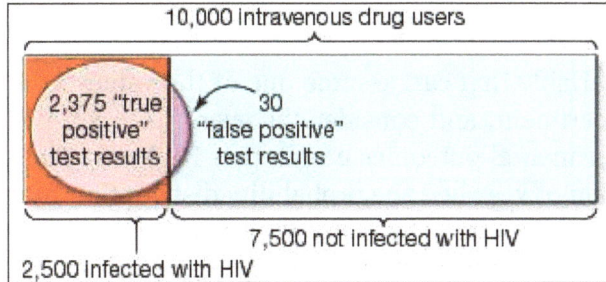

Bayes's theorem used for evaluating the accuracy of a medical testA hypothetical HIV test given to 10,000 intravenous drug users might produce 2,405 positive test results, which would include 2,375 "true positives" plus 30 "false positives. "Based on this experience, a physician would determine that the probability of a positive test result revealing an actual infection is 2,375 out of 2,405- an accuracy rate of 98.8 percent.

$$P(A_n | B) = \frac{P(A_n)P(B|A_n)}{\sum_i P(A_i)P(B|A_i)}.$$

As an example, suppose that two balls are drawn without replacement from an urn containing r red and b black balls. Let A be the event "red on the first draw" and B the event "red on the second draw." From the obvious relations $P(A) = r/(r + b) = 1 - P(A^c)$, $P(B|A) = (r - 1)/(r + b - 1)$, $P(B|A^c) = r/(r + b - 1)$, and Bayes's theorem, it follows that the probability of a red ball on the first draw given that the second one is known to be red equals $(r - 1)/(r + b - 1)$. A more interesting and important use of Bayes's theorem appears below in the discussion of subjective probabilities.

Random Variables, Distributions, Expectation and Variance

Random Variables

Usually it is more convenient to associate numerical values with the outcomes of an experiment than to work directly with a nonnumerical description such as "red ball on the first draw." For example, an outcome of the experiment of drawing n balls with replacement from an urn containing black and red balls is an n-tuple that tells us whether a red or a black ball was drawn on each of the draws. This n-tuple is conveniently represented by an n-tuple of ones and zeros, where the appearance of a one in the kth position indicates that a red ball was drawn on the kth draw. A quantity of particular interest is the number of red balls drawn, which is just the sum of the entries in this numerical description of the experimental outcome. Mathematically a rule that associates with every element of a given set a unique real number is called a "(real-valued) function." In the history of statistics and probability, real-valued functions defined on a sample space have traditionally been called "random variables." Thus, if a sample space S has the generic element e, the outcome of an experiment, then a random variable is a real-valued function $X = X(e)$. Customarily one omits the argument e in the notation for a random variable. For the experiment of drawing balls from an urn containing black and red balls, R, the number of red balls drawn, is a random variable. A particularly useful random variable is 1[A], the indicator variable of the event A, which equals 1 if A occurs and 0 otherwise. A "constant" is a trivial random variable that always takes the same value regardless of the outcome of the experiment.

Probability Distribution

Suppose X is a random variable that can assume one of the values $x_1, x_2,..., x_m$, according to the outcome of a random experiment, and consider the event $\{X = x_i\}$, which is a shorthand notation for the set of all experimental outcomes e such that $X(e) = x_i$. The probability of this event, $P\{X = x_i\}$, is itself a function of x_i, called the probability distribution function of X. Thus, the distribution of the random variable R defined in the preceding section is the function of i = 0, 1,..., n given in the binomial equation. Introducing the notation $f(x_i) = P\{X = x_i\}$, one sees from the basic properties of probabilities that:

$$f(x_i) \geq 0 \text{ for all } i, \ \sum_i f(x_i) = 1,$$

and

$$P\{a < X \leq b\} = \sum_{a<x_j\leq b} f(x_i),$$

for any real numbers a and b. If Y is a second random variable defined on the same sample space as X and taking the values $y_1, y_2,..., yn$, the function of two variables $h(x_i, y_j) = P\{X = x_i, Y = y_j\}$ is called the joint distribution of X and Y. Since $\{X = x_i\} = \cup_j\{X = x_i, Y = y_j\}$, and this union consists of disjoint events in the sample space,

$$f(x_i) = \sum_j h(x_i, y_j), \text{ for all } i.$$

Often f is called the marginal distribution of X to emphasize its relation to the joint distribution of X and Y. Similarly, $g(y_j) = \Sigma_i h(x_i, y_j)$ is the (marginal) distribution of Y. The random variables X and Y are defined to be independent if the events $\{X = x_i\}$ and $\{Y = y_j\}$ are independent for all i and j—i.e., if $h(x_i, y_j) = f(x_i)g(y_j)$ for all i and j. The joint distribution of an arbitrary number of random variables is defined similarly.

Suppose two dice are thrown. Let X denote the sum of the numbers appearing on the two dice, and let Y denote the number of even numbers appearing. The possible values of X are 2, 3,..., 12, while the possible values of Y are 0, 1, 2. Since there are 36 possible outcomes for the two dice, the accompanying table giving the joint distribution $h(i, j)$ (i = 2, 3,..., 12; j = 0, 1, 2) and the marginal distributions $f(i)$ and $g(j)$ is easily computed by direct enumeration.

For more complex experiments, determination of a complete probability distribution usually requires a combination of theoretical analysis and empirical experimentation and is often very difficult. Consequently, it is desirable to describe a distribution insofar as possible by a small number of parameters that are comparatively easy to evaluate and interpret. The most important are the mean and the variance. These are both defined in terms of the "expected value" of a random variable.

Expected Value

Given a random variable X with distribution f, the expected value of X, denoted $E(X)$, is defined by $E(X) = \Sigma_i x_i f(x_i)$. In words, the expected value of X is the sum of each of the possible values of X multiplied by the probability of obtaining that value. The expected value of X is also called the mean of the distribution f. The basic property of E is that of linearity: if X and Y are random variables and if a and b are constants, then $E(aX + bY) = aE(X) + bE(Y)$. To see why this is true, note that $aX + bY$ is itself a random variable, which assumes the values $ax_i + by_j$ with the probabilities $h(x_i, y_j)$. Hence,

$$E(aX + bY) = \sum_{i,j}(ax_i + by_j)h(x_i, y_j)$$
$$= a\sum_{i,j}x_i h(x_i, y_j) + b\sum_{i,j} y_j h(x_i, y_j).$$

If the first sum on the right-hand side is summed over j while holding i fixed, by equation above the result is,

$$f(x_i) = \sum_j h(x_i, y_j), \text{ for all } i.$$

$$\sum_i x_i f(x_i),$$

which by definition is $E(X)$. Similarly, the second sum equals $E(Y)$.

If 1[A] denotes the "indicator variable" of A—i.e., a random variable equal to 1 if A occurs and equal to 0 otherwise—then $E\{1[A]\} = 1 \times P(A) + 0 \times P(A^c) = P(A)$. This shows that the concept of expectation includes that of probability as a special case.

As an illustration, consider the number R of red balls in n draws with replacement from an urn containing a proportion p of red balls. From the definition and the binomial distribution of R,

$$E(R) = \sum_i i\binom{n}{i}p^i q^{n=i},$$

which can be evaluated by algebraic manipulation and found to equal np. It is easier to use the representation $R = 1[A_1] + \cdots + 1[A_n]$, where A_k denotes the event "the kth draw results in a red ball." Since $E\{1[A_k]\} = p$ for all k, by linearity $E(R) = E\{1[A_1]\} + \cdots + E\{1[A_n]\} = np$. This argument illustrates the principle that one can often compute the expected value of a random variable without first computing its distribution. For another example, suppose n balls are dropped at random into n boxes. The number of empty boxes, Y, has the representation $Y = 1[B_1] + \cdots + 1[B_n]$, where B_k is the event that "the kth box is empty." Since the kth box is empty if and only if each of the n balls went into one of the other $n - 1$ boxes, $P(B_k) = [(n - 1)/n]^n$ for all k, and consequently $E(Y) = n(1 - 1/n)^n$. The exact distribution of Y is very complicated, especially if n is large.

Many probability distributions have small values of $f(x_i)$ associated with extreme (large or small) values of x_i and larger values of $f(x_i)$ for intermediate x_i. For example, both marginal distributions in the table are symmetrical about a midpoint that has relatively high probability, and the probability of other values decreases as one moves away from the midpoint. Insofar as a distribution $f(x_i)$ follows this kind of pattern, one can interpret the mean of f as a rough measure of location of the bulk of the probability distribution, because in the defining sum the values x_i associated with large values of $f(x_i)$ more or less define the centre of the distribution. In the extreme case, the expected value of a constant random variable is just that constant.

Variance

It is also of interest to know how closely packed about its mean value a distribution is. The most important measure of concentration is the variance, denoted by $Var(X)$ and defined by $Var(X) = E\{[X - E(X)]^2\}$. By linearity of expectations, one has equivalently $Var(X) = E(X^2) - \{E(X)\}^2$. The standard deviation of X is the square root of its variance. It has a more direct interpretation than the variance because it is in the same units as X. The variance of a constant random variable is 0. Also, if c is a constant, $Var(cX) = c^2 Var(X)$.

There is no general formula for the expectation of a product of random variables. If the random variables X and Y are independent, $E(XY) = E(X)E(Y)$. This can be used to show that, if $X_1, \ldots,$ Xn are independent random variables, the variance of the sum $X_1 + \cdots + Xn$ is just the sum of the individual variances, $Var(X_1) + \cdots + Var(X_n)$. If the Xs have the same distribution and are independent, the variance of the average $(X_1 + \cdots + X_n)/n$ is $Var(X_1)/n$. Equivalently, the standard deviation of $(X_1 + \cdots + X_n)/n$ is the standard deviation of X_1 divided by $\sqrt{n}$. This quantifies the intuitive notion that the average of repeated observations is less variable than the individual observations. More precisely, it says that the variability of the average is inversely proportional to the square root of the number of observations. This result is tremendously important in problems of statistical inference.

Consider again the binomial distribution given by equation $\binom{n}{i} = \dfrac{r^i b^{n-i}}{(r+b)^n} = \binom{n}{i} p^i q^{n-i}$ $(i = 0, 1, 2, \ldots, n)$. As in the calculation of the mean value, one can use the definition combined with some algebraic manipulation to show that, if R has the binomial distribution, then $Var(R) = npq$. From the representation $R = 1[A_1] + \cdots + 1[A_n]$ defined above, and the observation that the events A_k are independent and have the same probability, it follows that:

$$Var(R) = Var\{1[A_1]\} + \ldots + nVar\{1[A_n]\} = Var\{1[A_1]\}.$$

Moreover,

$$Var\{1[A_1]\} = E\{1[A_1]\}^2 - E\{1[A_n]\}^2 = p - p^2 = pq,$$

so $Var(R) = npq$.

The conditional distribution of Y given $X = x_i$ is defined by:

$$P\{Y = y_j \mid X = x_i\} = \frac{h(x_i y_j)}{f(x_i)}$$

and the conditional expectation of Y given $X = xi$ is:

$$E(Y|X = x_i) = \sum_j \frac{y_j\, h(x_i, y_j)}{f(x_{i)}}$$

One can regard $E(Y|X)$ as a function of X; since X is a random variable, this function of X must itself be a random variable. The conditional expectation $E(Y|X)$ considered as a random variable has its own (unconditional) expectation $E\{E(Y|X)\}$, which is calculated by multiplying equation above by $f(xi)$ and summing over i to obtain the important formula:

$$E\{E(Y|X)\} = E(Y)$$

Properly interpreted, equation above is a generalization of the law of total probability.

For a simple example of the use of equation above, recall the problem of the gambler's ruin and let $e(x)$ denote the expected duration of the game if Peter's fortune is initially equal to x. The reasoning leading to equation $Q(x) = pQ(x+1) + qQ(x-1)$ in conjunction with equation above shows that $e(x)$ satisfies the equations $e(x) = 1 + pe(x + 1) + qe(x - 1)$ for $x = 1, 2,..., m - 1$ with the boundary conditions $e(0) = e(m) = 0$. The solution for $p \neq 1/2$ is rather complicated; for $p = 1/2$, $e(x) = x(m - x)$.

An Alternative Interpretation of Probability

In ordinary conversation the word probability is applied not only to variable phenomena but also to propositions of uncertain veracity. The truth of any proposition concerning the outcome of an experiment is uncertain before the experiment is performed. Many other uncertain propositions cannot be defined in terms of repeatable experiments. An individual can be uncertain about the truth of a scientific theory, a religious doctrine, or even about the occurrence of a specific historical event when inadequate or conflicting eyewitness accounts are involved. Using probability as a measure of uncertainty enlarges its domain of application to phenomena that do not meet the requirement of repeatability. The concomitant disadvantage is that probability as a measure of uncertainty is subjective and varies from one person to another.

According to one interpretation, to say that someone has subjective probability p that a proposition is true means that for any integers r and b with $r/(r + b) < p$, if that individual is offered an opportunity to bet the same amount on the truth of the proposition or on "red in a single draw" from an urn containing r red and b black balls, he prefers the first bet, while, if $r/(r + b) > p$, he prefers the second bet.

An important stimulus to modern thought about subjective probability has been an attempt to understand decision making in the face of incomplete knowledge. It is assumed that an individual, when faced with the necessity of making a decision that may have different consequences depending on situations about which he has incomplete knowledge, can express his personal preferences and uncertainties in a way consistent with certain axioms of rational behaviour. It can then be deduced that the individual has a utility function, which measures the value to him of each course of action when each of the uncertain possibilities is the true one, and a "subjective probability distribution," which expresses quantitatively his beliefs about the uncertain situations. The individual's optimal decision is the one that maximizes his expected utility with respect to his subjective probability. The concept of utility goes back at least to Daniel Bernoulli (Jakob Bernoulli's nephew) and was developed in the 20th century by John von Neumannand Oskar Morgenstern, Frank P. Ramsey, and Leonard J. Savage, among others. Ramsey and Savage stressed the importance of subjective probability as a concomitant ingredient of decision making in the face of uncertainty. An alternative approach to subjective probability without the use of utility theory was developed by Bruno de Finetti.

The mathematical theory of probability is the same regardless of one's interpretation of the concept, although the importance attached to various results can depend very much on the interpretation. In particular, in the theory and applications of subjective probability, Bayes's theorem plays an important role.

For example, suppose that an urn contains N balls, r of which are red and $b = N - r$ of which are black, but r (hence b) is unknown. One is permitted to learn about the value of r by performing the experiment of drawing with replacement n balls from the urn. Suppose also that one has a subjective probability distribution giving the probability $f(r)$ that the number of red balls is in fact r where $f(0) + \cdots + f(N) = 1$. This distribution is called an a priori distribution because it is specified prior to the experiment of drawing balls from the urn. The binomial distribution is now a conditional probability, given the value of r. Finally, one can use Bayes's theorem to find the conditional probability that the unknown number of red balls in the urn is r, given that the number of red balls drawn from the urn is i. The result is

$$\frac{f(r)r^i b^{n-i}}{\sum_{r0=0}^{n} f(r_0)r_0^i b_0^{n-i}}, \text{ where } b_0 = N - r_0.$$

This distribution, derived by using Bayes's theorem to combine the a priori distribution with the conditional distribution for the outcome of the experiment, is called the a posteriori distribution.

The virtue of this calculation is that it makes possible a probability statement about the composition of the urn, which is not directly observable, in terms of observable data, from the composition of the sample taken from the urn. The weakness, as indicated above, is that different people may choose different subjective probabilities for the composition of the urn a priori and hence reach different conclusions about its composition a posteriori.

To see how this idea might apply in practice, consider a simple urn model of opinion polling to predict which of two candidates will win an election. The red balls in the urn are identified with voters who will vote for candidate A and the black balls with those voting for candidate B. Choosing a sample from the electorate and asking their preferences is a well-defined random experiment, which in theory and in practice is repeatable. The composition of the urn is uncertain and is not the result of a well-defined random experiment. Nevertheless, to the extent that a vote for a candidate is a vote for a political party, other elections provide information about the content of the urn, which, if used judiciously, should be helpful in supplementing the results of the actual sample to make a prediction. Exactly how to use this information is a difficult problem in which individual judgment plays an important part. One possibility is to incorporate the prior information into an a priori distribution about the electorate, which is then combined via Bayes's theorem with the outcome of the sample and summarized by an a posteriori distribution.

The Law of Large Numbers, the Central Limit Theorem and the Poisson Approximation

The relative frequency interpretation of probability is that if an experiment is repeated a large number of times under identical conditions and independently, then the relative frequency with which an event A actually occurs and the probability of A should be approximately the same. A mathematical expression of this interpretation is the law of large numbers. This theorem says that if $X_1, X_2,..., X_n$ are independent random variables having a common distribution with mean μ, then for any number $\varepsilon > 0$, no matter how small, as $n \to \infty$,

$$P\{\left|n^{-1})(X_1 +...+ X_n)-\mu\right| < \varepsilon \to 1$$

The law of large numbers was first proved by Jakob Bernoulli in the special case where X_k is 1 or 0 according as the kth draw (with replacement) from an urn containing r red and b black balls is red or black. Then $E(X_k) = r/(r + b)$, and the last equation says that the probability that "the difference between the empirical proportion of red balls in n draws and the probability of red on a single draw is less than ε" converges to 1 as n becomes infinitely large.

Insofar as an event which has probability very close to 1 is practically certain to happen, this result justifies the relative frequency interpretation of probability. Strictly speaking, however, the justification is circular because the probability in the above equation, which is very close to but not equal to 1, requires its own relative frequency interpretation. Perhaps it is better to say that the weak law of large numbers is consistent with the relative frequency interpretation of probability.

The following simple proof of the law of large numbers is based on Chebyshev's inequality, which illustrates the sense in which the variance of a distribution measures how the distribution is dispersed about its mean. If X is a random variable with distribution f and mean μ, then by definition $\text{Var}(X) = \Sigma_i(x_i - \mu)^2 f(x_i)$. Since all terms in this sum are positive, the sum can only decrease if some of the terms are omitted. Suppose one omits all terms with $|x_i - \mu| < b$, where b is an arbitrary given number. Each term remaining in the sum has a factor of the form $(x_i - \mu)^2$, which is greater than or equal to b^2. Hence, $\text{Var}(X) \geq b^2 \Sigma' f(x_i)$, where the prime on the summation sign indicates that only terms with $|x_i - \mu| \geq b$ are included in the sum. Chebyshev's inequality is this expression rewritten as:

$$P\{\left|X - \mu\right| \geq b\} \leq \frac{Var(X)}{b^2}.$$

This inequality can be applied to the complementary event of that appearing in $P\{\left|n^{-1})(X_1 +...+ X_n)-\mu\right| < \varepsilon \to 1$, with $b = \varepsilon$. The Xs are independent and have the same distribution, $E[n^{-1}(X_1 +\cdots+ X_n)] = \mu$ and $\text{Var}[(X_1 +\cdots+ X_n)/n] = \text{Var}(X_1)/n$, so that:

$$P\left\{\left|\frac{(X_1 +...+ X_n)}{n} - \mu\right| \geq \varepsilon\right\} \geq 1 - \frac{Var(X_1)}{n\varepsilon^2}.$$

This not only proves $P\{\left|n^{-1})(X_1 +...+ X_n)-\mu\right| < \varepsilon \to 1$, but it also says quantitatively how large n should be in order that the empirical average, $n^{-1}(X_1 +\cdots+ X_n)$, approximate its expectation to any required degree of precision.

Suppose, for example, that the proportion p of red balls in an urn is unknown and is to be estimated by the empirical proportion of red balls in a sample of size n drawn from the urn with replacement. Chebyshev's inequality with $X_k = 1\{$red ball on the kth draw$\}$ implies that, in order that the observed proportion be within ε of the true proportion p with probability at least 0.95, it suffices that n be at least $20 \times \mathrm{Var}(X_1)/\varepsilon^2$. Since $\mathrm{Var}(X_1) = p(1-p) \le 1/4$ for all p, for $\varepsilon = 0.03$ it suffices that n be at least 5,555. It is shown below that this value of n is much larger than necessary, because Chebyshev's inequality is not sufficiently precise to be useful in numerical calculations.

Although Jakob Bernoulli did not know Chebyshev's inequality, the inequality he derived was also imprecise, and, perhaps because of his disappointment in not having a quantitatively useful approximation, he did not publish the result during his lifetime. It appeared in 1713, eight years after his death.

The Central Limit Theorem

The desired useful approximation is given by the central limit theorem, which in the special case of the binomial distribution was first discovered by Abraham de Moivre about 1730. Let $X_1,..., X_n$ be independent random variables having a common distribution with expectation μ and variance σ^2. The law of large numbers implies that the distribution of the random variable $\bar{X}_n = n^{-1}(X_1 +\cdots+ X_n)$ is essentially just the degenerate distribution of the constant μ, because $E(\bar{X}_n n) = \mu$ and $\mathrm{Var}(\bar{X}_n) = \sigma^2/n \to 0$ as $n \to \infty$. The standardized random variable $(\bar{X}_n - \mu)/(\sigma/\sqrt{n})$ has mean 0 and variance 1. The central limit theorem gives the remarkable result that, for any real numbers a and b, as $n \to \infty$,

$$P\left\{a < \frac{(\bar{X}_n - \mu)}{(\sigma/n1/2)}\right\} \to G(b) - G(a),$$

where

$$G(z) = \frac{1}{\sqrt{2\pi}} \int_{-\infty}^{z} \exp\left(\frac{-t^2}{2}\right) dt.$$

Thus, if n is large, the standardized average has a distribution that is approximately the same, regardless of the original distribution of the Xs. The equation also illustrates clearly the square root law: the accuracy of $\bar{X}_n$ as an estimator of μ is inversely proportional to the square root of the sample size n.

The right-hand side of this equation is the Poisson distribution. Its mean and variance are both equal to μt. Although the Poisson approximation is not comparable to the central limit theorem in importance, it nevertheless provides one of the basic building blocks in the theory of stochastic processes.

Infinite Sample Spaces and Axiomatic Probability

Infinite Sample Spaces

The experiments described in the preceding discussion involve finite sample spaces for the most part, although the central limit theorem and the Poisson approximation involve limiting operations and hence lead to integrals and infinite series. In a finite sample space, calculation of the probability of an event A is conceptually straightforward because the principle of additivity tells one to calculate the probability of a complicated event as the sum of the probabilities of the individual experimental outcomes whose union defines the event.

Experiments having a continuum of possible outcomes—for example, that of selecting a number at random from the interval $[r, s]$—involve subtle mathematical difficulties that were not satisfactorily resolved until the 20th century. If one chooses a number at random from $[r, s]$, the probability that the number falls in any interval $[x, y]$ must be proportional to the length of that interval; and, since the probability of the entire sample space $[r, s]$ equals 1, the constant of proportionality equals $1/(s - r)$. Hence, the probability of obtaining a number in the interval $[x, y]$ equals $(y - x)/(s - r)$. From this and the principle of additivity one can determine the probability of any event that can be expressed as a finite union of intervals. There are, however, very complicated sets having no simple relation to the intervals—e.g., the rational numbers—and it is not immediately clear what the probabilities of these sets should be. Also, the probability of selecting exactly the number x must be 0, because the set consisting of x alone is contained in the interval $[x, x + 1/n]$ for all n and hence must have no larger probability than $1/[n(s - r)]$, no matter how large n is. Consequently, it makes no sense to try to compute the probability of an event by "adding" the probabilities of the individual outcomes making up the event, because each individual outcome has probability 0.

A closely related experiment, although at first there appears to be no connection, arises as follows. Suppose that a coin is tossed n times, and let $Xk = 1$ or 0 according as the outcome of the kth toss is heads or tails. The weak law of large numbers given above says that a certain sequence of numbers—namely the sequence of probabilities and defined in terms of these n Xs—converges to 1 as $n \to \infty$. In order to formulate this result, it is only necessary to imagine that one can toss the coin n times and that this finite number of tosses can be arbitrarily large. In other words, there is a sequence of experiments, but each one involves a finite sample space. It is also natural to ask whether the sequence of random variables $(X_1 + \cdots + Xn)/n$ converges as $n \to \infty$. However, this question cannot even be formulated mathematically unless infinitely many Xs can be defined on the same sample space, which in turn requires that the underlying experiment involve an actual infinity of coin tosses.

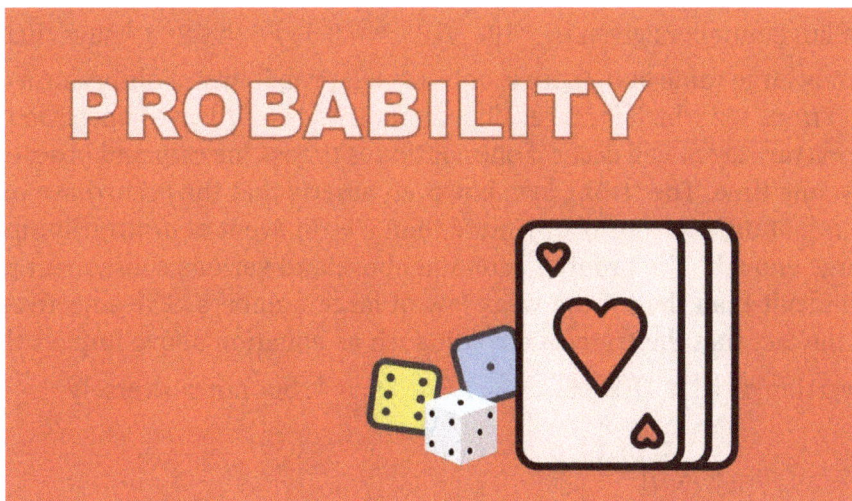

For the conceptual experiment of tossing a fair coin infinitely many times, the sequence of zeros and ones, $(X_1, X_2,...)$, can be identified with that real number that has the Xs as the coefficients of its expansion in the base 2, namely $X_1/2^1 + X_2/2^2 + X_3/2^3 +\cdots$. For example, the outcome of getting heads on the first two tosses and tails thereafter corresponds to the real number $1/2 + 1/4 + 0/8 +\cdots = 3/4$. (There are some technical mathematical difficulties that arise from the fact that some numbers have two representations. Obviously $1/2 = 1/2 + 0/4 +\cdots$, and the formula for the sum of an infinite geometric series shows that it also equals $0/2 + 1/4 + 1/8 +\cdots$. It can be shown that these difficulties do not pose a serious problem, and they are ignored in the subsequent discussion.) For any particular specification $i_1, i_2,...$, in of zeros and ones, the event $\{X_1 = i_1, X_2 = i_2,..., Xn = in\}$ must have probability $1/2n$ in order to be consistent with the experiment of tossing the coin only n times. Moreover, this event corresponds to the interval of real numbers $[i_1/2^1+ i_2/2^2 +\cdots+ in/2n, i_1/2^1 + i_2/2^2 +\cdots+ in/2n + 1/2n]$ of length $1/2n$, since any continuation Xn_{+1}, $Xn_{+2},...$ corresponds to a number that is at least 0 and at most $1/2n^{+1} + 1/2n^{+2} +\cdots = 1/2n$ by the formula for an infinite geometric series. It follows that the mathematical model for choosing a number at random from $[0, 1]$ and that of tossing a fair coin infinitely many times assign the same probabilities to all intervals of the form $[k/2n, 1/2n]$.

The Strong Law of Large Numbers

The mathematical relation between these two experiments was recognized in 1909 by the French mathematician Émile Borel, who used the then new ideas of measure theory to give a precise mathematical model and to formulate what is now called the strong law of large numbers for fair cointossing. His results can be described as follows. Let e denote a number chosen at random from $[0, 1]$, and let $X_k(e)$ be the kth coordinate in the expansion of e to the base 2. Then $X_1, X_2,...$ are an infinite sequence of independent random variables taking the values 0 or 1 with probability $1/2$ each. Moreover, the subset of $[0, 1]$ consisting of those e for which the sequence $n^{-1}[X_1(e) +\cdots+ X_n(e)]$ tends to $1/2$ as $n \to \infty$ has probability 1. Symbolically:

$$P\left\{\lim_{n\to\infty}[n^{-1}(X_1 +...+ X_n)] = \frac{1}{2}\right\} = 1.$$

The weak law of large numbers given in $P\{|n^{-1})(X_1 +...+ X_n) - \mu| < \varepsilon \to 1$ says that for any $\varepsilon > 0$, for each sufficiently large value of n, there is only a small probability of observing a deviation of $X_n = n^{-1}(X_1 +\cdots+ X_n)$ from $1/2$ which is larger than ε; nevertheless, it leaves open the possibility that sooner or later this rare event will occur if one continues to toss the coin and observe the sequence for a sufficiently long time. The strong law, however, asserts that the occurrence of even one value of X_k for $k \geq n$ that differs from $1/2$ by more than ε is an event of arbitrarily small probability provided n is large enough. The proof of equation above and various subsequent generalizations is much more difficult than that of the weak law of large numbers. The adjectives "strong" and "weak" refer to the fact that the truth of a result such as equation above implies the truth of the corresponding version of $P\{|n^{-1})(X_1 +...+ X_n) - \mu| < \varepsilon \to 1$, but not conversely.

$$P\left\{\lim_{n\to\infty}[n^{-1}(X_1 +...+ X_n)] = \frac{1}{2}\right\} = 1.$$

Measure Theory

During the two decades following 1909, measure theory was used in many concrete problems of probability theory, notably in the American mathematician Norbert Wiener's treatment of the mathematical theory of Brownian motion, but the notion that all problems of probability theory could be formulated in terms of measure is customarily attributed to the Soviet mathematician Andrey Nikolayevich Kolmogorov in 1933.

The fundamental quantities of the measure theoretic foundation of probability theory are the sample space S, which as before is just the set of all possible outcomes of an experiment, and a distinguished class M of subsets of S, called events. Unlike the case of finite S, in general not every subset of S is an event. The class M must have certain properties described below. Each event is assigned a probability, which means mathematically that a probability is a function P mapping M into the real numbers that satisfies certain conditions derived from one's physical ideas about probability.

The properties of M are as follows: (i) $S \in M$; (ii) if $A \in M$, then $Ac \in M$; (iii) if $A_1, A_2,... \in M$, then $A_1 \cup A_2 \cup \cdots \in M$. Recalling that M is the domain of definition of the probability P, one can interpret (i) as saying that $P(S)$ is defined, (ii) as saying that, if the probability of A is defined, then the probability of "not A" is also defined, and (iii) as saying that, if one can speak of the probability of each of a sequence of events An individually, then one can speak of the probability that at least one of the An occurs. A class of subsets of any set that has properties (i)–(iii) is called a σ-field. From these properties one can prove others. For example, it follows at once from (i) and (ii) that Ø (the empty set) belongs to the class M. Since the intersection of any class of sets can be expressed as the complement of the union of the complements of those sets (DeMorgan's law), it follows from (ii) and (iii) that, if $A_1, A_2,... \in M$, then $A_1 \cap A_2 \cap \cdots \in M$.

Given a set S and a σ-field M of subsets of S, a probability measure is a function P that assigns to each set $A \in M$ a nonnegative real number and that has the following two properties: (a) $P(S) = 1$ and (b) if $A_1, A_2,... \in M$ and $Ai \cap Aj = Ø$ for all $i \neq j$, then $P(A_1 \cup A_2 \cup \cdots) = P(A_1) + P(A_2) + \cdots$. Property (b) is called the axiom of countable additivity. It is clearly motivated by $P(A_1 \cup A_2 \cup \cdots \cup A_n) = P(A_1) + ... + P(A_n)$, which suffices for finite sample spaces because there are only finitely many events. In infinite sample spaces it implies, but is not implied by, $P(A_1 \cup A_2 \cup \cdots \cup A_n) = P(A_1) + ... + P(A_n)$. There is, however, nothing in one's intuitive notion of probability that requires the acceptance of this property. Indeed, a few mathematicians have developed probability theory with only the weaker axiom of finite additivity, but the absence of interesting models that fail to satisfy the axiom of countable additivity has led to its virtually universal acceptance.

To get a better feeling for this distinction, consider the experiment of tossing a biased coin having probability p of heads and $q = 1 - p$ of tails until heads first appears. To be consistent with the idea that the tosses are independent, the probability that exactly n tosses are required equals $qn^{-1}p$, since the first $n - 1$ tosses must be tails, and they must be followed by a head. One can imagine that this experiment never terminates—i.e., that the coin continues to turn up tails forever. By the axiom of countable additivity, however, the probability that heads occurs at some finite value of n equals $p + qp + q^2p + \cdots = p/(1 - q) = 1$, by the formula for the sum of an infinite geometric series. Hence, the probability that the experiment goes on forever equals 0. Similarly, one can compute the probability that the number of tosses is odd, as $p + q^2p + q^4p + \cdots = p/(1 - q^2) = 1/(1 + q)$. On the other hand, if only finite additivity were required, it would be possible to define the following admittedly bizarre probability. The sample space S is the set of all natural numbers, and the σ-field M is the class of all subsets of S. If an event A contains finitely many elements, $P(A) = 0$, and, if the complement of A contains finitely many elements, $P(A) = 1$. As a consequence of the deceptively innocuous axiom of choice (which says that, given any collection C of nonempty sets, there exists a rule for selecting a unique point from each set in C), one can show that many finitely additive probabilities consistent with these requirements exist. However, one cannot be certain what the probability of getting an odd number is, because that set is neither finite nor its complement finite, nor can it be expressed as a finite disjoint union of sets whose probability is already defined.

It is a basic problem, and by no means a simple one, to show that the intuitive notion of choosing a number at random from [0, 1], is consistent with the preceding definitions. Since the probability of an interval is to be its length, the class of events M must contain all intervals; but in order to be a σ-field it must contain other sets, many of which are difficult to describe in an elementary way. One

example is the event in $P\left\{\lim_{n\to\infty}[n^{-1}(X_1 + ... + X_n)] = \frac{1}{2}\right\} = 1.$, which must belong to M in order that

one can talk about its probability. Also, although it seems clear that the length of a finite disjoint union of intervals is just the sum of their lengths, a rather subtle argument is required to show that length has the property of countable additivity. A basic theorem says that there is a suitable σ-field containing all the intervals and a unique probability defined on this σ-field for which the probability of an interval is its length. The σ-field is called the class of Lebesgue-measurable sets, and the probability is called the Lebesgue measure, after the French mathematician and principal architect of measure theory, Henri-Léon Lebesgue.

In general, a σ-field need not be all subsets of the sample space S. The question of whether all subsets of [0, 1] are Lebesgue-measurable turns out to be a difficult problem that is intimately connected with the foundations of mathematics and in particular with the axiom of choice.

Probability Density Functions

For random variables having a continuum of possible values, the function that plays the same role as the probability distribution of a discrete random variable is called a probability density function. If the random variable is denoted by X, its probability density function f has the property that:

$$P\{a < X \le b\} = \int_a^b f(x)dx$$

for every interval (a, b); i.e., the probability that X falls in (a, b] is the area under the graph of f between a and b. For example, if X denotes the outcome of selecting a number at random from the interval [r, s], the probability density function of X is given by f(x) = 1/(s − r) for r < x < s and f(x) = 0 for x < r or x > s. The function F(x) defined by F(x) = P{X ≤ x} is called the distribution function, or cumulative distribution function, of X. If X has a probability density function f(x), the relation between f and F is F′(x) = f(x) or equivalently:

$$F(x) = \int_{-\infty}^{x} f(t)dt.$$

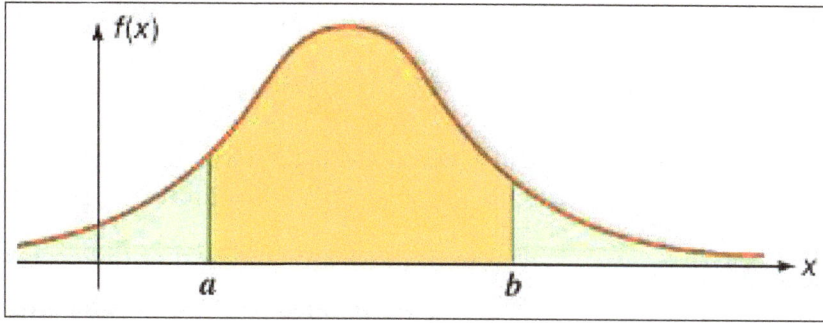

Probability density function.

The distribution function F of a discrete random variable should not be confused with its probability distribution f. In this case the relation between F and f is:

$$F(x) = \sum_{x_i \le x} f(x_i)$$

If a random variable X has a probability density function f(x), its "expectation" can be defined by:

$$E(X) = \int_{-\infty}^{\infty} xf(x)dx.$$

provided that this integral is convergent. It turns out to be simpler, however, not only to use Lebesgue's theory of measure to define probabilities but also to use his theory of integration to define expectation. Accordingly, for any random variable X, E(X) is defined to be the Lebesgue integral of X with respect to the probability measure P, provided that the integral exists. In this way it is possible to provide a unified theory in which all random variables, both discrete and continuous, can be treated simultaneously. In order to follow this path, it is necessary to restrict the class of those functions X defined on S that are to be called random variables, just as it was necessary to restrict the class of subsets of S that are called events. The appropriate restriction is that a random variable must be a measurable function. The definition is taken over directly from the Lebesgue theory of integration and will not be discussed here. It can be shown that, whenever X has a probability density function, its expectation (provided it exists) is given by equation above, which remains a useful formula for calculating E(X).

Some important probability density functions are the following:

$(i) \ Normal \ f(x) = (2\pi\sigma^2)^{-1} \exp\left[\dfrac{-(x-\mu)^2}{2\sigma^2}\right]$

$(ii) \ (-\infty < x < +\infty); \quad E(X) = \mu, Var(x), Var(X) = \mu^2$

$\quad Exponential : f(x) = \mu ex \exp(-\mu x)(0 \le x < +\infty),$

$\quad f(x) = 0(x < 0); \quad E(X) = \dfrac{1}{\mu}.$

$(iii) \ cauchy : f(x) = \dfrac{1}{[\pi(1+x^2)]}(-\infty < x < +\infty);$

$\quad E(X) \ does \ not \ exist.$

The cumulative distribution function of the normal distribution with mean 0 and variance 1 has already appeared as the function G defined following $P\left\{a < \dfrac{(\bar{X}n - \mu)}{(\sigma / n^{1/2})} \le b\right\} \to G(b) - G(a)$.

The law of large numbers and the central limit theoremcontinue to hold for random variables on infinite sample spaces. A useful interpretation of the central limit theorem stated formally in $P\left\{a < \dfrac{(\bar{X}n - \mu)}{(\sigma / n^{1/2})} \le b\right\} \to G(b) - G(a)$, is as follows: The probability that the average (or sum) of a large number of independent, identically distributed random variables with finite variance falls in an interval $(c_1, c_2]$ equals approximately the area between c_1 and c_2 underneath the graph of a normal density function chosen to have the same expectation and variance as the given average (or sum). The figureillustrates the normal approximation to the binomial distribution with $n = 10$ and $p = 1/2$.

The exponential distribution arises naturally in the study of the Poisson distribution introduced in $P\{N(t) = i\} = \dfrac{(\mu t)^i \exp(-\mu t)}{i!} (i = 0, 1, ...)$. If T_k denotes the time interval between the emission of the $k - $1st and kth particle, then $T_1, T_2,...$ are independent random variables having an exponential distribution with parameter μ. This is obvious for T_1 from the observation that $\{T_1 > t\} = \{N(t) = 0\}$. Hence, $P\{T_1 \le t\} = 1 - P\{N(t) = 0\} = 1 - \exp(-\mu t)$, and by differentiation one obtains the exponential density function.

The Cauchy distribution does not have a mean value or a variance, because the integral $E(X) = \int_{-\infty}^{\infty} xf(x)dx$ does not converge. As a result, it has a number of unusual properties. For example, if $X_1, X_2,..., X_n$ are independent random variables having a Cauchy distribution, the average $(X_1 + \cdots + X_n)/n$ also has a Cauchy distribution. The variability of the average is exactly the same as that of a single observation. Another random variable that does not have an expectation is the waiting time until the number of heads first equals the number of tails in tossing a fair coin.

Conditional Expectation and Least Squares Prediction

An important problem of probability theory is to predict the value of a future observation Y given knowledge of a related observation X (or, more generally, given several related observations X_1, $X_2,...$). Examples are to predict the future course of the national economy or the path of a rocket, given its present state.

Prediction is often just one aspect of a "control" problem. For example, in guiding a rocket, measurements of the rocket's location, velocity, and so on are made almost continuously; at each reading, the rocket's future course is predicted, and a control is then used to correct its future course. The same ideas are used to steer automatically large tankers transporting crude oil, for which even slight gains in efficiency result in large financial savings.

Given X, a predictor of Y is just a function $H(X)$. The problem of "least squares prediction" of Y given the observation X is to find that function $H(X)$ that is closest to Y in the sense that the mean square error of prediction, $E\{[Y - H(X)]^2\}$, is minimized. The solution is the conditional expectation $H(X) = E(Y|X)$.

In applications a probability model is rarely known exactly and must be constructed from a combination of theoretical analysis and experimental data. It may be quite difficult to determine the optimal predictor, $E(Y|X)$, particularly if instead of a single X a large number of predictor variables $X_1, X_2,...$ are involved. An alternative is to restrict the class of functions H over which one searches to minimize the mean square error of prediction, in the hope of finding an approximately optimal predictor that is much easier to evaluate. The simplest possibility is to restrict consideration to linear functions $H(X) = a + bX$. The coefficients a and b that minimize the restricted mean square prediction error $E\{(Y - a - bX)^2\}$ give the best linear least squares predictor. Treating this restricted mean square prediction error as a function of the two coefficients (a, b) and minimizing it by methods of the calculus yield the optimal coefficients: $\hat{b} = E\{[X - E(X)][Y - E(Y)]\}/\mathrm{Var}(X)$ and $\hat{a} = E(Y) - \hat{b}E(X)$. The numerator of the expression for $\hat{b}$ is called the covariance of X and Y and is denoted $\mathrm{Cov}(X, Y)$. Let $\hat{Y} = \hat{a} + \hat{b}X$ denote the optimal linear predictor. The mean square error of prediction is $E\{(Y - \hat{Y})^2\} = \mathrm{Var}(Y) - [\mathrm{Cov}(X, Y)]^2/\mathrm{Var}(X)$.

If X and Y are independent, then $\mathrm{Cov}(X, Y) = 0$, the optimal predictor is just $E(Y)$, and the mean square error of prediction is $\mathrm{Var}(Y)$. Hence, $|\mathrm{Cov}(X, Y)|$ is a measure of the value X has in predicting Y. In the extreme case that $[\mathrm{Cov}(X, Y)]^2 = \mathrm{Var}(X)\mathrm{Var}(Y)$, Y is a linear function of X, and the optimal linear predictor gives error-free prediction.

There is one important case in which the optimal mean square predictor actually is the same as the optimal linear predictor. If X and Y are jointly normally distributed, the conditional expectation of Y given X is just a linear function of X, and hence the optimal predictor and the optimal linear predictor are the same. The form of the bivariate normal distribution as well as expressions for the coefficients $\hat{a}$ and $\hat{b}$ and for the minimum mean square error of prediction were discovered by the English eugenicist Sir Francis Galton in his studies of the transmission of inheritable characteristics from one generation to the next. They form the foundation of the statistical technique of linear regression.

The Poisson Process and The Brownian Motion Process

The theory of stochastic processes attempts to build probability models for phenomena that evolve over time.

The Poisson Process

An important stochastic process described implicitly in the discussion of the Poisson approximation to the binomial distribution is the Poisson process. Modeling the emission of radioactive particles by an infinitely large number of tosses of a coin having infinitesimally small probability for heads on each toss led to the conclusion that the number of particles $N(t)$ emitted in the time interval $[0, t]$ has the Poisson distribution given in $P\{N(t) = i\} = \dfrac{(\mu t)^i \exp(-\mu t)}{i!}$ $(i = 0,1,...)$ with expectation μt. The primary concern of the theory of stochastic processes is not this marginal distribution of $N(t)$ at a particular time but rather the evolution of $N(t)$ over time. Two properties of the Poisson process that make it attractive to deal with theoretically are: (i) The times between emission of particles are independent and exponentially distributed with expected value $1/\mu$. (ii) Given that $N(t) = n$, the times at which the n particles are emitted have the same joint distribution as n points distributed independently and uniformly on the interval $[0, t]$.

As a consequence of property (i), a picture of the function $N(t)$ is very easily constructed. Originally $N(0) = 0$. At an exponentially distributed time T_1, the function $N(t)$ jumps from 0 to 1. It remains at 1 another exponentially distributed random time, T_2, which is independent of T_1, and at time $T_1 + T_2$ it jumps from 1 to 2, and so on.

Examples of other phenomena for which the Poisson process often serves as a mathematical model are the number of customers arriving at a counter and requesting service, the number of claims against an insurance company, or the number of malfunctions in a computer system. The importance of the Poisson process consists in (a) its simplicity as a test case for which the mathematical theory, and hence the implications, are more easily understood than for more realistic models and (b) its use as a building block in models of complex systems.

Brownian Motion Process

The most important stochastic process is the Brownian motion or Wiener process. It was first discussed by Louis Bachelier, who was interested in modeling fluctuations in prices in financial markets, and by Albert Einstein, who gave a mathematical model for the irregular motion of colloidal particles first observed by the Scottish botanist Robert Brown in 1827. The first mathematically rigorous treatment of this model was given by Wiener. Einstein's results led to an early, dramatic confirmation of the molecular theory of matter in the French physicist Jean Perrin's experiments to determine Avogadro's number, for which Perrin was awarded a Nobel Prize in 1926. Today somewhat different models for physical Brownian motion are deemed more appropriate than Einstein's, but the original mathematical model continues to play a central role in the theory and application of stochastic processes.

Let $B(t)$ denote the displacement (in one dimension for simplicity) of a colloidally suspended particle, which is buffeted by the numerous much smaller molecules of the medium in which it is suspended. This displacement will be obtained as a limit of a random walk occurring in discrete time as the number of steps becomes infinitely large and the size of each individual step infinitesimally small. Assume that at times $k\delta$, $k = 1, 2,...$, the colloidal particle is displaced a distance hX_k, where $X_1, X_2,...$ are +1 or −1 according as the outcomes of tossing a fair coin are heads or tails. By time t the particle has taken m steps, where m is the largest integer $\leq t/\delta$, and its displacement from its original position is $Bm(t) = h(X_1 + \cdots + Xm)$. The expected value of $Bm(t)$ is 0, and its variance is h^2m, or approximately h^2t/δ. Now suppose that $\delta \to 0$, and at the same time $h \to 0$ in such a way that the variance of $Bm(1)$ converges to some positive constant, σ^2. This means that m becomes infinitely large, and h is approximately $\sigma(t/m)^{1/2}$. It follows from the central limit theorem (equation 12) that $\lim P\{Bm(t) \leq x\} = G(x/\sigma t^{1/2})$, where $G(x)$ is the standard normal cumulative distribution function. The Brownian motion process $B(t)$ can be defined to be the limit in a certain technical sense of the $Bm(t)$ as $\delta \to 0$ and $h \to 0$ with $h^2/\delta \to \sigma^2$.

The process $B(t)$ has many other properties, which in principle are all inherited from the approximating random walk $Bm(t)$. For example, if (s_1, t_1) and (s_2, t_2) are disjoint intervals, the increments $B(t_1) - B(s_1)$ and $B(t_2) - B(s_2)$ are independent random variables that are normally distributed with expectation 0 and variances equal to $\sigma^2(t_1 - s_1)$ and $\sigma^2(t_2 - s_2)$, respectively.

Einstein took a different approach and derived various properties of the process $B(t)$ by showing that its probability density function, $g(x, t)$, satisfies the diffusion equation $\partial g/\partial t = D\partial^2 g/\partial x^2$, where $D = \sigma^2/2$. The important implication of Einstein's theory for subsequent experimental research was that he identified the diffusion constant D in terms of certain measurable properties of the particle (its radius) and of the medium (its viscosity and temperature), which allowed one to make predictions and hence to confirm or reject the hypothesized existence of the unseen molecules that were assumed to be the cause of the irregular Brownian motion. Because of the beautiful blend of mathematical and physical reasoning involved, a brief summary of the successor to Einstein's model is given below.

Unlike the Poisson process, it is impossible to "draw" a picture of the path of a particle undergoing mathematical Brownian motion. Wiener showed that the functions $B(t)$ are continuous, as one expects, but nowhere differentiable. Thus, a particle undergoing mathematical Brownian motion does not have a well-defined velocity, and the curve $y = B(t)$ does not have a well-defined tangent at any value of t. To see why this might be so, recall that the derivative of $B(t)$, if it exists, is the limit as $h \rightarrow 0$ of the ratio $[B(t + h) - B(t)]/h$. Since $B(t + h) - B(t)$ is normally distributed with mean 0 and standard deviation $h^{1/2}\sigma$, in very rough terms $B(t + h) - B(t)$ can be expected to equal some multiple (positive or negative) of $h^{1/2}$. But the limit as $h \rightarrow 0$ of $h^{1/2}/h = 1/h^{1/2}$ is infinite. A related fact that illustrates the extreme irregularity of $B(t)$ is that in every interval of time, no matter how small, a particle undergoing mathematical Brownian motion travels an infinite distance. Although these properties contradict the commonsense idea of a function—and indeed it is quite difficult to write down explicitly a single example of a continuous, nowhere-differentiable function—they turn out to be typical of a large class of stochastic processes, called diffusion processes, of which Brownian motion is the most prominent member. Especially notable contributions to the mathematical theory of Brownian motion and diffusion processes were made by Paul Lévy and William Feller during the years 1930–60.

A more sophisticated description of physical Brownian motion can be built on a simple application of Newton's second law: $F = ma$. Let $V(t)$ denote the velocity of a colloidal particle of mass m. It is assumed that:

$$mdV(t) = -fV(t)dt + dA(t).$$

The quantity f retarding the movement of the particle is due to friction caused by the surrounding medium. The term $dA(t)$ is the contribution of the very frequent collisions of the particle with unseen molecules of the medium. It is assumed that f can be determined by classical fluid mechanics, in which the molecules making up the surrounding medium are so many and so small that the medium can be considered smooth and homogeneous. Then by Stokes's law, for a spherical particle in a gas, $f = 6\pi a\eta$, where a is the radius of the particle and η the coefficient of viscosity of the medium. Hypotheses concerning $A(t)$ are less specific, because the molecules making up the surrounding medium cannot be observed directly. For example, it is assumed that, for $t \neq s$, the infinitesimal random increments $dA(t) = A(t + dt) - A(t)$ and $A(s + ds) - A(s)$ caused by collisions of the particle with molecules of the surrounding medium are independent random variables having distributions with mean 0 and unknown variances $\sigma^2\, dt$ and $\sigma^2\, ds$ and that $dA(t)$ is independent of $dV(s)$ for $s < t$.

The differential equation above has the solution:

$$mdV(t) = -fV(t)dt + dA.$$

$$V(t) = V(0)\exp(-\beta t) + m^{-1}\int_0^t \exp[\beta(t - s)]dA(s),$$

where $\beta = f/m$. From this equation and the assumed properties of $A(t)$, it follows that $E[V^2(t)] \to \sigma^2/(2mf)$ as $t \to \infty$. Now assume that, in accordance with the principle of equipartition of energy, the steady-state average kinetic energy of the particle, $m \lim_{t \to \infty} E[V^2(t)]/2$, equals the average kinetic energy of the molecules of the medium. According to the kinetic theory of an ideal gas, this is $RT/2N$, where R is the ideal gas constant, T is the temperature of the gas in kelvins, and N is Avogadro's number, the number of molecules in one gram molecular weight of the gas. It follows that the unknown value of σ^2 can be determined: $\sigma^2 = 2RTf/N$.

If one also assumes that the functions $V(t)$ are continuous, which is certainly reasonable from physical considerations, it follows by mathematical analysis that $A(t)$ is a Brownian motion process as defined above. This conclusion poses questions about the meaning of the initial $mdV(t) = -fV(t)dt + dA(t)$, because for mathematical Brownian motion the term $dA(t)$ does not exist in the usual sense of a derivative. Some additional mathematical analysis shows that the stochastic differential $mdV(t) = -fV(t)dt + dA(t)$ and its solution equation below have a precise mathematical interpretation. The process $V(t)$ is called the Ornstein-Uhlenbeck process, after the physicists Leonard Salomon Ornstein and George Eugene Uhlenbeck. The logical outgrowth of these attempts to differentiate and integrate with respect to a Brownian motion process is the Ito (named for the Japanese mathematician Itō Kiyosi) stochastic calculus, which plays an important role in the modern theory of stochastic processes.

The displacement at time t of the particle whose velocity is given by equation below:

$$X(t) - X(0) \int_0^t V(u)du = \beta^{-1}V(0)[1 - \exp(-\beta t)] + f^{-1}A(t) - f^{-1}\int_0^t \exp[-\beta(t-u)]dA(u).$$

For t large compared with β, the first and third terms in this expression are small compared with the second. Hence, $X(t) - X(0)$ is approximately equal to $A(t)/f$, and the mean square displacement, $E\{[X(t) - X(0)]^2\}$, is approximately $\sigma^2/f^2 = RT/(3\pi a\eta N)$. These final conclusions are consistent with Einstein's model, although here they arise as an approximation to the model obtained from equation above. Since it is primarily the conclusions that have observational consequences, there are essentially no new experimental implications. However, the analysis arising directly out of Newton's second law, which yields a process having a well-defined velocity at each point, seems more satisfactory theoretically than Einstein's original model.

Stochastic Processes

A stochastic process is a family of random variables $X(t)$ indexed by a parameter t, which usually takes values in the discrete set $T = \{0, 1, 2,...\}$ or the continuous set $T = [0, +\infty)$. In many cases t represents time, and $X(t)$ is a random variable observed at time t. Examples are the Poisson process, the Brownian motion process, and the Ornstein-Uhlenbeck process described in the preceding section. Considered as a totality, the family of random variables $\{X(t), t \in T\}$ constitutes a "random function."

Stationary Processes

The mathematical theory of stochastic processes attempts to define classes of processes for which a unified theory can be developed. The most important classes are stationary processes and Markov processes. A stochastic process is called stationary if, for all n, $t_1 < t_2 < \cdots < tn$, and $h > 0$, the joint distribution of $X(t_1 + h),..., X(tn + h)$ does not depend on h. This means that in effect there is no origin on the time axis; the stochastic behaviour of a stationary process is the same no matter when the process is observed. A sequence of independent identically distributed random variables is an example of a stationary process. A rather different example is defined as follows: $U(0)$ is uniformly distributed on $[0, 1]$; for each $t = 1, 2,..., U(t) = 2U(t - 1)$ if $U(t - 1) \leq 1/2$, and $U(t) = 2U(t - 1) - 1$ if $U(t - 1) > 1/2$. The marginal distributions of $U(t)$, $t = 0, 1,...$ are uniformly distributed on $[0, 1]$, but, in contrast to the case of independent identically distributed random variables, the entire sequence can be predicted from knowledge of $U(0)$. A third example of a stationary process is:

$$X(t) = \sum_k ck[Y_k \cos(\theta_{kt}) + Z_k \sin(\theta_{kt}),$$

where the Ys and Zs are independent normally distributed random variables with mean 0 and unit variance, and the cs and θs are constants. Processes of this kind can be useful in modeling seasonal or approximately periodic phenomena.

A remarkable generalization of the strong law of large numbers is the ergodic theorem: if $X(t)$, $t = 0, 1,...$ for the discrete case or $0 \leq t < \infty$ for the continuous case, is a stationary process such that $E[X(0)]$ is finite, then with probability 1 the average:

$$s^1 \sum_{t=0}^{s-1} X(t), \text{if t is discrete, or } s^{-1} \int_0^s X(t)dt,$$

if t is continuous, converges to a limit as $s \rightarrow \infty$. In the special case that t is discrete and the Xs are independent and identically distributed, the strong law of large numbers is also applicable and shows that the limit must equal $E\{X(0)\}$. However, the example that $X(0)$ is an arbitrary random variable and $X(t) \equiv X(0)$ for all $t > 0$ shows that this cannot be true in general. The limit does equal $E\{X(0)\}$ under an additional rather technical assumption to the effect that there is no subset of the state space, having probability strictly between 0 and 1, in which the process can get stuck and never escape. This assumption is not fulfilled by the example $X(t) \equiv X(0)$ for all t, which gets stuck immediately at its initial value. It is satisfied by the sequence $U(t)$ defined above, so by the ergodic theorem the average of these variables converges to $1/2$ with probability 1. The ergodic theorem was first conjectured by the American chemist J. Willard Gibbs in the early 1900s in the context of statistical mechanics and was proved in a corrected, abstract formulation by the American mathematician George David Birkhoff in 1931.

Markovian Processes

A stochastic process is called Markovian (after the Russian mathematician Andrey Andreyevich Markov) if at any time t the conditional probability of an arbitrary future event given the entire past of the process—i.e., given $X(s)$ for all $s \leq t$—equals the conditional probability of that future event given only $X(t)$. Thus, in order to make a probabilistic statement about the future behaviour of a Markov process, it is no more helpful to know the entire history of the process than it is to know only its current state. The conditional distribution of $X(t + h)$ given $X(t)$ is called the transition probability of the process. If this conditional distribution does not depend on t, the process is said to have "stationary" transition probabilities. A Markov process with stationary transition probabilities may or may not be a stationary process in the sense of the preceding paragraph. If $Y_1, Y_2,...$ are independent

random variables and $X(t) = Y_1 + \cdots + Yt$, the stochastic process $X(t)$ is a Markov process. Given $X(t) = x$, the conditional probability that $X(t + h)$ belongs to an interval (a, b) is just the probability that $Yt_{+1} + \cdots + Yt_{+}h$ belongs to the translated interval $(a - x, b - x)$; and because of independence this conditional probability would be the same if the values of $X(1),\ldots, X(t - 1)$ were also given. If the Ys are identically distributed as well as independent, this transition probability does not depend on t, and then $X(t)$ is a Markov process with stationary transition probabilities. Sometimes $X(t)$ is called a random walk, but this terminology is not completely standard. Since both the Poisson process and Brownian motion are created from random walks by simple limiting processes, they, too, are Markov processes with stationary transition probabilities. The Ornstein-Uhlenbeck process defined as the $X(t) - X(0) \int_0^t V(u)du = \beta^{-1}V(0)[1 - \exp(-\beta t)] + f^{-1}A(t) - f^{-1}\int_0^t \exp[-\beta(t-u)]dA(u)$
to the stochastic differential $mdV(t) = -fV(t)dt + dA(t)$ is also a Markov process with stationary transition probabilities.

The Ornstein-Uhlenbeck process and many other Markov processes with stationary transition probabilities behave like stationary processes as $t \to \infty$. Roughly speaking, the conditional distribution of $X(t)$ given $X(0) = x$ converges as $t \to \infty$ to a distribution, called the stationary distribution, that does not depend on the starting value $X(0) = x$. Moreover, with probability 1, the proportion of time the process spends in any subset of its state space converges to the stationary probability of that set; and, if $X(0)$ is given the stationary distribution to begin with, the process becomes a stationary process. The Ornstein-Uhlenbeck process defined in

$$X(t) - X(0) \int_0^t V(u)du = \beta^{-1}V(0)[1 - \exp(-\beta t)] + f^{-1}A(t) - f^{-1}\int_0^t \exp[-\beta(t-u)]dA(u)$$ is stationary

if $V(0)$ has a normal distribution with mean 0 and variance $\sigma^2/(2mf)$.

At another extreme are absorbing processes. An example is the Markov process describing Peter's fortune during the game of gambler's ruin. The process is absorbed whenever either Peter or Paul is ruined. Questions of interest involve the probability of being absorbed in one state rather than another and the distribution of the time until absorption occurs.

The Ehrenfest Model of Diffusion

The Ehrenfest model of diffusion (named after the Austrian Dutch physicist Paul Ehrenfest) was proposed in the early 1900s in order to illuminate the statistical interpretation of the second law of thermodynamics, that the entropy of a closed system can only increase. Suppose N molecules of a gas are in a rectangular container divided into two equal parts by a permeable membrane. The state of the system at time t is $X(t)$, the number of molecules on the left-hand side of the membrane. At each time $t = 1, 2,\ldots$ a molecule is chosen at random (i.e., each molecule has probability $1/N$ to be chosen) and is moved from its present location to the other side of the membrane. Hence, the system evolves according to the transition probability $p(i, j) = P\{X(t + 1) = j | X(t) = i\}$, where,

$$p(i, i+1) = 1 - \frac{i}{N}, \quad p(i, i-1) = \frac{i}{N},$$
$$p(i, j) = 0 \quad \text{for } j \neq i+1, i-1.$$

The long run behaviour of the Ehrenfest process can be inferred from general theorems about Markov processes in discrete time with discrete state space and stationary transition probabilities.

Let $T(j)$ denote the first time $t \geq 1$ such that $X(t) = j$ and set $T(j) = \infty$ if $X(t) \neq j$ for all t. Assume that for all states i and j it is possible for the process to go from i to j in some number of steps—i.e., $P\{T(j) < \infty | X(0) = i\} > 0$. If the equations,

$$Q(j) = \sum_i Q(i) p(i.j)$$

have a solution $Q(j)$ that is a probability distribution—i.e., $Q(j) \geq 0$, and $\Sigma Q(j) = 1$— then that solution is unique and is the stationary distribution of the process. Moreover, $Q(j) = 1/E\{T(j)|X(0) = j\}$; and, for any initial state j, the proportion of time t that $X(t) = i$ converges with probability 1 to $Q(i)$.

For the special case of the Ehrenfest process, assume that N is large and $X(0) = 0$. According to the deterministic prediction of the second law of thermodynamics, the entropy of this system can only increase, which means that $X(t)$ will steadily increase until half the molecules are on each side of the membrane. Indeed, according to the stochastic model described above, there is overwhelming probability that $X(t)$ does increase initially. However, because of random fluctuations, the system occasionally moves from configurations having large entropy to those of smaller entropy and eventually even returns to its starting state, in defiance of the second law of thermodynamics.

The accepted resolution of this contradiction is that the length of time such a system must operate in order that an observable decrease of entropy may occur is so enormously long that a decrease could never be verified experimentally. To consider only the most extreme case, let T denote the first time $t \geq 1$ at which $X(t) = 0$—i.e., the time of first return to the starting configuration having all molecules on the right-hand side of the membrane. It can be verified by substitution in $Q(j) = \sum_i Q(i) p(i.j)$ that the stationary distribution of the Ehrenfest model is the binomial distribution:

$$Q(i) = \binom{n}{j} 2^{-N},$$

and hence $E(T) = 2N$. For example, if N is only 100 and transitions occur at the rate of 10^6 per second, $E(T)$ is of the order of 10^{15} years. Hence, on the macroscopic scale, on which experimental measurements can be made, the second law of thermodynamics holds.

The Symmetric Random Walk

A Markov process that behaves in quite different and surprising ways is the symmetric random walk. A particle occupies a point with integer coordinates in d-dimensional Euclidean space. At each time $t = 1, 2,...$ it moves from its present location to one of its $2d$ nearest neighbours with equal probabilities $1/(2d)$, independently of its past moves. For $d = 1$ this corresponds to moving a step to the right or left according to the outcome of tossing a fair coin. It may be shown that for $d = 1$ or 2 the particle returns with probability 1 to its initial position and hence to every possible position infinitely many times, if the random walk continues indefinitely. In three or more dimensions, at any time t the number of possible steps that increase the distance of the particle from the origin is much larger than the number decreasing the distance, with the result that the particle eventually moves away from the origin and never returns. Even in one or two dimensions, although the particle eventually returns to its initial position, the expected waiting time until it returns is infinite, there is no stationary distribution, and the proportion of time the particle spends in any state converges to 0.

Queuing Models

The simplest service system is a single-server queue, where customers arrive, wait their turn, are served by a single server, and depart. Related stochastic processes are the waiting time of the nth customer and the number of customers in the queue at time t. For example, suppose that customers arrive at times $0 = T_0 < T_1 < T_2 < \cdots$ and wait in a queue until their turn. Let Vn denote the service time required by the nth customer, $n = 0, 1, 2,...$, and set $Un = Tn - Tn_{-1}$. The waiting time, Wn, of the nth customer satisfies the relation $W_0 = 0$ and, for $n \geq 1$, $Wn = \max(0, Wn_{-1} + Vn_{-1} - Un)$. Observe that the nth customer must wait for the same length of time as the $(n - 1)$th customer plus the service time of the $(n - 1)$th customer minus the time between the arrival of the $(n - 1)$th and nth customer, during which the $(n - 1)$th customer is already waiting but the nth customer is not. An exception occurs if this quantity is negative, and then the waiting time of the nth customer is 0. Various assumptions can be made about the input and service mechanisms. One possibility is that customers arrive according to a Poisson process and their service times are independent, identically distributed random variables that are also independent of the arrival process. Then, in terms of $Yn = Vn_{-1} - Un$, which are independent, identically distributed random variables, the recursive relation defining Wn becomes $Wn = \max(0, Wn_{-1} + Yn)$. This process is a Markov process. It is often called a random walk with reflecting barrier at 0, because it behaves like a random walk whenever it is positive and is pushed up to be equal to 0 whenever it tries to become negative. Quantities of interest are the mean and variance of the waiting time of the nth customer and, since these are very difficult to determine exactly, the mean and variance of the stationary distribution. More realistic queuing models try to accommodate systems with several servers and different classes of customers, who are served according to certain priorities. In most cases it is impossible to give a mathematical analysis of the system, which must be simulated on a computer in order to obtain numerical results. The insights gained from theoretical analysis of simple cases can be helpful in performing these simulations. Queuing theory had its origins in attempts to understand traffic in telephone systems. Present-day research is stimulated, among other things, by problems associated with multiple-user computer systems.

Reflecting barriers arise in other problems as well. For example, if $B(t)$ denotes Brownian motion, then $X(t) = B(t) + ct$ is called Brownian motion with drift c. This model is appropriate for Brownian motion of a particle under the influence of a constant force field such as gravity. One can add a reflecting barrier at 0 to account for reflections of the Brownian particle off the bottom of its container. The result is a model for sedimentation, which for $c < 0$ in the steady state as $t \to \infty$ gives a statistical derivation of the law of pressure as a function of depth in an isothermal atmosphere. Just as ordinary Brownian motion can be obtained as the limit of a rescaled random walk as the number of steps becomes very large and the size of individual steps small, Brownian motion with a reflecting barrier at 0 can be obtained as the limit of a rescaled random walk with reflection at 0. In this way, Brownian motion with a reflecting barrier plays a role in the analysis of queuing systems. In fact, in modern probability theory one of the most important uses of Brownian motion and other diffusion processes is as approximations to more complicated stochastic processes. The exact mathematical description of these approximations gives remarkable generalizations of the central limit theorem from sequences of random variables to sequences of random functions.

Insurance Risk Theory

The ruin problem of insurance risk theory is closely related to the problem of gambler's ruin described earlier and, rather surprisingly, to the single-server queue as well. Suppose the amount of capital at time t in one portfolio of an insurance company is denoted by $X(t)$. Initially $X(0) = x > 0$. During each unit of time, the portfolio receives an amount $c > 0$ in premiums. At random times claims are made against the insurance company, which must pay the amount $Vn > 0$ to settle the nth claim. If $N(t)$ denotes the number of claims made in time t. Then,

$$X(t) = x + ct - \sum_{1}^{N(t)} V_n,$$

provided that this quantity has been positive at all earlier times $s < t$. At the first time $X(t)$ becomes negative, however, the portfolio is ruined. A principal problem of insurance risk theory is to find the probability of ultimate ruin. If one imagines that the problem of gambler's ruin is modified so that Peter's opponent has an infinite amount of capital and can never be ruined, then the probability that Peter is ultimately ruined is similar to the ruin probability of insurance risk theory. In fact, with the artificial assumptions that (i) $c = 1$, (ii) time proceeds by discrete units, say $t = 1, 2,...$, (iii) Vn is identically equal to 2 for all n, and (iv) at each time t a claim occurs with probability p or does not occur with probability q independently of what occurs at other times, then the process $X(t)$ is the same stochastic process as Peter's fortune, which is absorbed if it ever reaches the state 0. The probability of Peter's ultimate ruin against an infinitely rich adversary is easily obtained by taking the limit of,

$$Q(x) = \frac{\left(\frac{q}{p}\right)^x - \left(\frac{q}{p}\right)^m}{1 - \left(\frac{q}{p}\right)^m} \left(p \neq \tfrac{1}{2}\right) = 1 - \frac{x}{m}\left(p = \tfrac{1}{2}\right)$$

as $m \to \infty$. The answer is $(q/p)x$ if $p > q$—i.e., the game is favourable to Peter—and 1 if $p \leq q$. More interesting assumptions for the insurance risk problem are that the number of claims $N(t)$ is a Poisson process and the sizes of the claims $V_1, V_2,...$ are independent, identically distributed positive random variables. Rather surprisingly, under these assumptions the probability of ultimate ruin as a function of the initial fortune x is exactly the same as the stationary probability that the waiting time in the single-server queue with Poisson input exceeds x. Unfortunately, neither problem is easy to solve exactly, although there is a very good approximate solution originally derived by the Swedish mathematician Harald Cramér.

Martingale Theory

As a final example, it seems appropriate to mention one of the dominant ideas of modern probability theory, which at the same time springs directly from the relation of probability to games of chance. Suppose that $X_1, X_2,...$ is any stochastic process and, for each $n = 0, 1,..., fn = fn(X_1,..., Xn)$ is a (Borel-measurable) function of the indicated observations. The new stochastic process fn is called a martingale if $E(fn|X_1,..., Xn_{-1}) = fn_{-1}$ for every value of $n > 0$ and all values of $X_1,..., Xn_{-1}$. If the sequence of Xs are outcomes in successive trials of a game of chance and fn is the fortune of a gambler after the nth trial, then the martingale condition says that the game is absolutely fair in the sense that, no matter what the past history of the game, the gambler's conditional expected fortune after one more trial is exactly equal to his present fortune. For example, let $X_0 = x$, and for $n \geq 1$ let Xn equal 1 or -1 according as a coin having probability p of heads and $q = 1 - p$ of tails turns up heads or tails on the nth toss. Let $Sn = X_0 + \cdots + Xn$. Then $fn = Sn - n(p - q)$ and $fn = (q/p)Sn$ are martingales. One of the basic results of martingale theory is that, if the gambler is free to quit the game at any time using any strategy whatever, provided only that this strategy does not foresee the future, then the game remains fair. This means that, if N denotes the stopping time at which the gambler's strategy tells him to quit the game, so that his final fortune is fN, then:

$$E(fN \mid f0) = f0.$$

Strictly speaking, this result is not true without some additional conditions that must be verified for any particular application. To see how efficiently it works, consider once again the problem of gambler's ruin and let N be the first value of n such that $Sn = 0$ or m; i.e., N denotes the random time at which ruin first occurs and the game ends. In the case $p = 1/2$, application of $E(fN \mid f0) = f0$ to the martingale $fn = Sn$, together with the observation that $fN =$ either 0 or m, yields the equalities $x = f_0 = E(fN \mid f_0 = x) = m[1 - Q(x)]$, which can be immediately solved to give the answer in,

$$Q(x) = \frac{\left(\frac{q}{p}\right)^x - \left(\frac{q}{p}\right)^m}{1 - \left(\frac{q}{p}\right)^m} \left(p \neq \tfrac{1}{2}\right) = 1 - \frac{x}{m} \left(p = \tfrac{1}{2}\right)$$

For $p \neq 1/2$, one uses the martingale $fn = (q/p)Sn$ and similar reasoning to obtain,

$$\left(\frac{q}{p}\right)^x = E(f_N \mid f0 = x) = 1Q(x) + \left(\frac{q}{p}\right)^m [1 - Q(x)],$$

from which the first equation,

$$Q(x) = \frac{\left(\frac{q}{p}\right)^x - \left(\frac{q}{p}\right)^m}{1 - \left(\frac{q}{p}\right)^m} \left(p \neq \tfrac{1}{2}\right) = 1 - \frac{x}{m} \left(p = \tfrac{1}{2}\right)$$

easily follows. The expected duration of the game is obtained by a similar argument.

A particularly beautiful and important result is the martingale convergence theorem, which implies that a nonnegative martingale converges with probability 1 as $n \to \infty$. This means that, if a gambler's successive fortunes form a (nonnegative) martingale, they cannot continue to fluctuate indefinitely but must approach some limiting value.

Statistical Theory

The theory of statistics provides a basis for the whole range of techniques, in both study design and data analysis, that are used within applications of statistics. The theory covers approaches to statistical-decision problems and to statistical inference, and the actions and deductions that satisfy the basic principles stated for these different approaches. Within a given approach, statistical theory gives ways of comparing statistical procedures; it can find a best possible procedure within a given context for given statistical problems, or can provide guidance on the choice between alternative procedures.

Apart from philosophical considerations about how to make statistical inferences and decisions, much of statistical theory consists of mathematical statistics, and is closely linked to probability theory, to utility theory, and to optimization.

Scope

Statistical theory provides an underlying rationale and provides a consistent basis for the choice of methodology used in applied statistics.

Modelling

Statistical models describe the sources of data and can have different types of formulation corresponding to these sources and to the problem being studied. Such problems can be of various kinds:

- Sampling from a finite population.

- Measuring observational error and refining procedures.

- Studying statistical relations.

Statistical models, once specified, can be tested to see whether they provide useful inferences for new data sets. Testing a hypothesis using the data that was used to specify the model is a fallacy, according to the natural science of Bacon and the scientific method of Peirce.

Data Collection

Statistical theory provides a guide to comparing methods of data collection, where the problem is to generate informative data using optimization and randomization while measuring and controlling for observational error. Optimization of data collection reduces the cost of data while satisfying statistical goals, while randomization allows reliable inferences. Statistical theory provides a basis for good data collection and the structuring of investigations in the topics of:

- Design of experiments to estimate treatment effects, to test hypotheses, and to optimize responses.

- Survey sampling to describe populations.

Summarising Data

The task of summarising statistical data in conventional forms (also known as descriptive statistics) is considered in theoretical statistics as a problem of defining what aspects of statistical samples need to be described and how well they can be described from a typically limited sample of data. Thus the problems theoretical statistics considers include:

- Choosing summary statistics to describe a sample.

- Summarising probability distributions of sample data while making limited assumptions about the form of distribution that may be met.

- Summarising the relationships between different quantities measured on the same items with a sample.

Interpreting Data

Besides the philosophy underlying statistical inference, statistical theory has the task of considering the types of questions that data analysts might want to ask about the problems they are studying and of providing data analytic techniques for answering them. Some of these tasks are:

- Summarising populations in the form of a fitted distribution or probability density function.

- Summarising the relationship between variables using some type of regression analysis.

- Providing ways of predicting the outcome of a random quantity given other related variables.

- Examining the possibility of reducing the number of variables being considered within a problem (the task of Dimension reduction).

When a statistical procedure has been specified in the study protocol, then statistical theory provides well-defined probability statements for the method when applied to all populations that could have arisen from the randomization used to generate the data. This provides an objective way of estimating parameters, estimating confidence intervals, testing hypotheses, and selecting the best. Even for observational data, statistical theory provides a way of calculating a value that can be used to interpret a sample of data from a population, it can provide a means of indicating how well that value is determined by the sample, and thus a means of saying corresponding values derived for different populations are as different as they might seem; however, the reliability of inferences from post-hoc observational data is often worse than for planned randomized generation of data.

Applied Statistical Inference

Statistical theory provides the basis for a number of data-analytic approaches that are common across scientific and social research. Interpreting data is done with one of the following approaches:

- Estimating parameters.

- Providing a range of values instead of a point estimate.

- Testing statistical hypotheses.

Many of the standard methods for those approaches rely on certain statistical assumptions (made in the derivation of the methodology) actually holding in practice. Statistical theory studies the consequences of departures from these assumptions. In addition it provides a range of robust statistical techniques that are less dependent on assumptions, and it provides methods checking whether particular assumptions are reasonable for a given data set.

Decision Theory

Decision theory is the study of an agent's choices. Decision theory can be broken into two branches: normative decision theory, which analyzes the outcomes of decisions or determines the optimal decisions given constraints and assumptions, and descriptive decision theory, which analyzes *how* agents actually make the decisions they do.

Decision theory is closely related to the field of game theory and is an interdisciplinary topic, studied by economists, statisticians, psychologists, biologists, political and other social scientists, philosophers, and computer scientists.

Empirical applications of this rich theory are usually done with the help of statistical and econometric methods.

Normative and Descriptive

Normative decision theory is concerned with identification of optimal decisions where optimality is often determined by considering an ideal decision maker who is able to compute with perfect accuracy and is in some sense fully rational. The practical application of this prescriptive approach (how people *ought to* make decisions) is called decision analysis and is aimed at finding tools, methodologies, and software (decision support systems) to help people make better decisions.

In contrast, positive or descriptive decision theory is concerned with describing observed behaviors often under the assumption that the decision-making agents are behaving under some consistent rules. These rules may, for instance, have a procedural framework (e.g. Amos Tversky's elimination by aspects model) or an axiomatic framework, reconciling the Von Neumann-Morgenstern axioms with behavioral violations of the expected utility hypothesis, or they may explicitly give a functional form for time-inconsistent utility functions (e.g. Laibson's quasi-hyperbolic discounting).

The prescriptions or predictions about behavior that positive decision theory produces allow for further tests of the kind of decision-making that occurs in practice. In recent decades, there has also been increasing interest in what is sometimes called "behavioral decision theory" and contributing to a re-evaluation of what useful decision-making requires.

Types of Decisions

Choice under Uncertainty

The area of choice under uncertainty represents the heart of decision theory. Known from the 17th century, the idea of expected value is that, when faced with a number of actions, each of which could give rise to more than one possible outcome with different probabilities, the rational procedure is to identify all possible outcomes, determine their values (positive or negative) and the probabilities that will result from each course of action, and multiply the two to give an "expected value", or the average expectation for an outcome; the action to be chosen should be the one that gives rise to the highest total expected value. In 1738, Daniel Bernoulli published an influential paper entitled *Exposition of a New Theory on the Measurement of Risk*, in which he uses the St. Petersburg paradox to show that expected value theory must be normatively wrong. He gives an example in which a Dutch merchant is trying to decide whether to insure a cargo being sent from Amsterdam to St Petersburg in winter. In his solution, he defines a utility function and computes expected utility rather than expected financial value.

In the 20th century, interest was reignited by Abraham Wald's 1939 paper pointing out that the two central procedures of sampling-distribution-based statistical-theory, namely hypothesis testing and parameter estimation, are special cases of the general decision problem. Wald's paper renewed and synthesized many concepts of statistical theory, including loss functions, risk functions, admissible decision rules, antecedent distributions, Bayesian procedures, and minimax procedures. The phrase "decision theory" itself was used in 1950 by E. L. Lehmann.

The revival of subjective probability theory, from the work of Frank Ramsey, Bruno de Finetti, Leonard Savage and others, extended the scope of expected utility theory to situations where subjective probabilities can be used. At the time, von Neumann and Morgenstern's theory of expected utility proved that expected utility maximization followed from basic postulates about rational behavior.

The work of Maurice Allais and Daniel Ellsberg showed that human behavior has systematic and sometimes important departures from expected-utility maximization. The prospect theory of Daniel Kahneman and Amos Tversky renewed the empirical study of economic behavior with less emphasis on rationality presuppositions. Kahneman and Tversky found three regularities – in actual human decision-making, "losses loom larger than gains"; persons focus more on *changes* in their utility-states than they focus on absolute utilities; and the estimation of subjective probabilities is severely biased by anchoring.

Intertemporal Choice

Intertemporal choice is concerned with the kind of choice where different actions lead to outcomes that are realised at different points in time. If someone received a windfall of several thousand dollars, they could spend it on an expensive holiday, giving them immediate pleasure, or they could invest it in a pension scheme, giving them an income at some time in the future. What is the optimal thing to do? The answer depends partly on factors such as the expected rates of interest and inflation, the person's life expectancy, and their confidence in the pensions industry. However even with all those factors taken into account, human behavior again deviates greatly from the predictions of prescriptive decision theory, leading to alternative models in which, for example, objective interest rates are replaced by subjective discount rates.

Interaction of Decision Makers

Some decisions are difficult because of the need to take into account how other people in the situation will respond to the decision that is taken. The analysis of such social decisions is more often treated under the label of game theory, rather than decision theory, though it involves the same mathematical methods. From the standpoint of game theory, most of the problems treated in decision theory are one-player games (or the one player is viewed as playing against an impersonal background situation). In the emerging field of socio-cognitive engineering, the research is especially focused on the different types of distributed decision-making in human organizations, in normal and abnormal/emergency/crisis situations.

Complex Decisions

Other areas of decision theory are concerned with decisions that are difficult simply because of their complexity, or the complexity of the organization that has to make them. Individuals making decisions may be limited in resources or are boundedly rational (have finite time or intelligence); in such cases the issue, more than the deviation between real and optimal behaviour, is the difficulty of determining the optimal behaviour in the first place. One example is the model of economic growth and resource usage developed by the Club of Rome to help politicians make real-life decisions in complex situations. Decisions are also affected by whether options are framed together or separately; this is known as the distinction bias. In 2011, Dwayne Rosenburgh explored and showed how decision theory can be applied to complex decisions that arise in areas such as wireless communications.

Heuristics

Heuristics in decision-making is the ability of making decisions based on unjustified or routine thinking. While quicker than step-by-step processing, heuristic thinking is also more likely to involve fallacies or inaccuracies. The main use for heuristics in our daily routines is to decrease the amount of evaluative thinking we perform when making simple decisions, making them instead based on unconscious rules and focusing on some aspects of the decision, while ignoring others. One example of a common and erroneous thought process that arises through heuristic thinking is the Gambler's Fallacy — believing that an isolated random event is affected by previous isolated random events. For example, if a coin is flipped to tails for a couple of turns, it still has the same probability of doing so; however it seems more likely, intuitively, for it to roll heads soon. This happens because, due to routine thinking, one disregards the probability and concentrates on the ratio of the outcomes, meaning that one expects that in the long run the ratio of flips should be half for each outcome. Another example is that decision-makers may be biased towards preferring moderate alternatives to extreme ones; the *Compromise Effect* operates under a mindset that the most moderate option carries the most benefit. In an incomplete information scenario, as in most daily decisions, the moderate option will look more appealing than either extreme, independent of the context, based only on the fact that it has characteristics that can be found at either extreme.

Alternatives

A highly controversial issue is whether one can replace the use of probability in decision theory by other alternatives.

Probability Theory

Advocates for the use of probability theory point to:

- The work of Richard Threlkeld Cox for justification of the probability axioms,

- The Dutch book paradoxes of Bruno de Finetti as illustrative of the theoretical difficulties that can arise from departures from the probability axioms, and

- The complete class theorems, which show that all admissible decision rules are equivalent to the Bayesian decision rule for some utility function and some prior distribution (or for the limit of a sequence of prior distributions). Thus, for every decision rule, either the rule may be reformulated as a Bayesian procedure (or a limit of a sequence of such), or there is a rule that is sometimes better and never worse.

Alternatives to Probability Theory

The proponents of fuzzy logic, possibility theory, quantum cognition, Dempster–Shafer theory, and info-gap decision theory maintain that probability is only one of many alternatives and point to many examples where non-standard alternatives have been implemented with apparent success; notably, probabilistic decision theory is sensitive to assumptions about the probabilities of various events, while non-probabilistic rules such as minimax are robust, in that they do not make such assumptions.

Ludic Fallacy

A general criticism of decision theory based on a fixed universe of possibilities is that it considers the "known unknowns", not the "unknown unknowns": it focuses on expected variations, not on unforeseen events, which some argue have outsized impact and must be considered – significant events may be "outside model". This line of argument, called the ludic fallacy, is that there are inevitable imperfections in modeling the real world by particular models, and that unquestioning reliance on models blinds one to their limits.

Estimation Theory

Estimation theory is a branch of statistics that deals with estimating the values of parameters based on measured empirical data that has a random component. The parameters describe an underlying physical setting in such a way that their value affects the distribution of the measured data. An estimator attempts to approximate the unknown parameters using the measurements.

In estimation theory, two approaches are generally considered:

- The probabilistic approach assumes that the measured data is random with probability distribution dependent on the parameters of interest.

- The set-membership approach assumes that the measured data vector belongs to a set which depends on the parameter vector.

For example, it is desired to estimate the proportion of a population of voters who will vote for a particular candidate. That proportion is the parameter sought; the estimate is based on a small random sample of voters. Alternatively, it is desired to estimate the probability of a voter voting for a particular candidate, based on some demographic features, such as age.

Or, for example, in radar the aim is to find the range of objects (airplanes, boats, etc.) by analyzing the two-way transit timing of received echoes of transmitted pulses. Since the reflected pulses are unavoidably embedded in electrical noise, their measured values are randomly distributed, so that the transit time must be estimated.

As another example, in electrical communication theory, the measurements which contain information regarding the parameters of interest are often associated with a noisy signal.

Basics:

For a given model, several statistical "ingredients" are needed so the estimator can be implemented. The first is a statistical sample – a set of data points taken from a random vector (RV) of size N. Put into a vector,

$$\mathbf{x} = \begin{bmatrix} x[0] \\ x[1] \\ \vdots \\ x[N-1] \end{bmatrix}.$$

Secondly, there are M parameters,

$$\theta = \begin{bmatrix} \theta_1 \\ \theta_2 \\ \vdots \\ \theta_M \end{bmatrix},$$

whose values are to be estimated. Third, the continuous probability density function (pdf) or its discrete counterpart, the probability mass function (pmf), of the underlying distribution that generated the data must be stated conditional on the values of the parameters:

$$p(\mathbf{x} \mid \theta)$$

It is also possible for the parameters themselves to have a probability distribution (e.g., Bayesian statistics). It is then necessary to define the Bayesian probability:

$$\pi(\theta)$$

After the model is formed, the goal is to estimate the parameters, with the estimates commonly denoted $\hat{\theta}$, where the "hat" indicates the estimate.

One common estimator is the minimum mean squared error (MMSE) estimator, which utilizes the error between the estimated parameters and the actual value of the parameters:

$$\mathbf{e} = \hat{\theta} - \theta$$

as the basis for optimality. This error term is then squared and the expected value of this squared value is minimized for the MMSE estimator.

Estimators

Commonly used estimators (estimation methods) and topics related to them include:

- Maximum likelihood estimators.

- Bayes estimators.

- Method of moments estimators.

- Cramér–Rao bound.

- Least squares.

- Minimum mean squared error (MMSE), also known as Bayes least squared error (BLSE).

- Maximum a posteriori (MAP).

- Minimum variance unbiased estimator (MVUE).

- Nonlinear system identification.

- Best linear unbiased estimator (BLUE).

- Unbiased estimators.

- Particle filter.

- Markov chain Monte Carlo (MCMC).

- Kalman filter, and its various derivatives.

- Wiener filter.

Unknown Constant in Additive White Gaussian Noise

Consider a received discrete signal, $x[n]$, of N independent samples that consists of an unknown constant A with additive white Gaussian noise (AWGN) $w[n]$ with known variance σ^2 (i.e., $\mathcal{N}(0,\sigma^2)$). Since the variance is known then the only unknown parameter is A.

The model for the signal is then

$$x[n] = A + w[n] \quad n = 0,1,\ldots,N-1$$

Two possible (of many) estimators for the parameter A are:

- $\hat{A}_1 = x[0]$,

- $\hat{A}_2 = \dfrac{1}{N}\sum_{n=0}^{N-1} x[n]$ which is the sample mean.

Both of these estimators have a mean of A, which can be shown through taking the expected value of each estimator:

$$E\left[\hat{A}_1\right] = E[x[0]] = A$$

and

$$E\left[\hat{A}_2\right] = E\left[\frac{1}{N}\sum_{n=0}^{N-1}x[n]\right] = \frac{1}{N}\left[\sum_{n=0}^{N-1}E[x[n]]\right] = \frac{1}{N}[NA] = A$$

At this point, these two estimators would appear to perform the same. However, the difference between them becomes apparent when comparing the variances,

$$\text{var}\left(\hat{A}_1\right) = \text{var}\left(x[0]\right) = \sigma^2$$

and

$$\text{var}\left(\hat{A}_2\right) = \text{var}\left(\frac{1}{N}\sum_{n=0}^{N-1}x[n]\right) \overset{\text{independence}}{=} \frac{1}{N^2}\left[\sum_{n=0}^{N-1}\text{var}(x[n])\right] = \frac{1}{N^2}[N\sigma^2] = \frac{\sigma^2}{N}$$

It would seem that the sample mean is a better estimator since its variance is lower for every $N > 1$.

Maximum Likelihood

Continuing the example using the maximum likelihood estimator, the probability density function (pdf) of the noise for one sample $w[n]$ is,

$$p(w[n]) = \frac{1}{\sigma\sqrt{2\pi}}\exp\left(-\frac{1}{2\sigma^2}w[n]^2\right)$$

and the probability of $x[n]$ becomes ($x[n]$ can be thought of a $\mathcal{N}(A,\sigma^2)$)

$$p(x[n]; A) = \frac{1}{\sigma\sqrt{2\pi}}\exp\left(-\frac{1}{2\sigma^2}(x[n]-A)^2\right)$$

By independence, the probability of x becomes,

$$p(\mathbf{x}; A) = \prod_{n=0}^{N-1}p(x[n]; A) = \frac{1}{\left(\sigma\sqrt{2\pi}\right)^N}\exp\left(-\frac{1}{2\sigma^2}\sum_{n=0}^{N-1}(x[n]-A)^2\right)$$

Taking the natural logarithm of the pdf,

$$\ln p(\mathbf{x}; A) = -N\ln\left(\sigma\sqrt{2\pi}\right) - \frac{1}{2\sigma^2}\sum_{n=0}^{N-1}(x[n]-A)^2$$

and the maximum likelihood estimator is,

$$\hat{A} = \arg\max \ln p(\mathbf{x}; A)$$

Taking the first derivative of the log-likelihood function,

$$\frac{\partial}{\partial A} \ln p(\mathbf{x}; A) = \frac{1}{\sigma^2} \left[\sum_{n=0}^{N-1} (x[n] - A) \right] = \frac{1}{\sigma^2} \left[\sum_{n=0}^{N-1} x[n] - NA \right]$$

and setting it to zero,

$$0 = \frac{1}{\sigma^2} \left[\sum_{n=0}^{N-1} x[n] - NA \right] = \sum_{n=0}^{N-1} x[n] - NA$$

This results in the maximum likelihood estimator,

$$\hat{A} = \frac{1}{N} \sum_{n=0}^{N-1} x[n]$$

which is simply the sample mean. From this example, it was found that the sample mean is the maximum likelihood estimator for N samples of a fixed, unknown parameter corrupted by AWGN.

Cramér–Rao Lower Bound

To find the Cramér–Rao lower bound (CRLB) of the sample mean estimator, it is first necessary to find the Fisher information number,

$$\mathcal{I}(A) = \mathrm{E}\left(\left[\frac{\partial}{\partial A} \ln p(\mathbf{x}; A) \right]^2 \right) = -\mathrm{E}\left[\frac{\partial^2}{\partial A^2} \ln p(\mathbf{x}; A) \right]$$

and copying from above,

$$\frac{\partial}{\partial A} \ln p(\mathbf{x}; A) = \frac{1}{\sigma^2} \left[\sum_{n=0}^{N-1} x[n] - NA \right]$$

Taking the second derivative,

$$\frac{\partial^2}{\partial A^2} \ln p(\mathbf{x}; A) = \frac{1}{\sigma^2} (-N) = \frac{-N}{\sigma^2}$$

and finding the negative expected value is trivial since it is now a deterministic constant,

$$-\mathrm{E}\left[\frac{\partial^2}{\partial A^2} \ln p(\mathbf{x}; A) \right] = \frac{N}{\sigma^2}$$

Finally, putting the Fisher information into:

$$\text{var}\left(\hat{A}\right) \geq \frac{1}{\mathcal{I}}$$

results in:

$$\text{var}\left(\hat{A}\right) \geq \frac{\sigma^2}{N}$$

Comparing this to the variance of the sample mean (determined previously) shows that the sample mean is *equal to* the Cramér–Rao lower bound for all values of N and A. In other words, the sample mean is the (necessarily unique) efficient estimator, and thus also the minimum variance unbiased estimator (MVUE), in addition to being the maximum likelihood estimator.

Maximum of a Uniform Distribution

One of the simplest non-trivial examples of estimation is the estimation of the maximum of a uniform distribution. It is used as a hands-on classroom exercise and to illustrate basic principles of estimation theory. Further, in the case of estimation based on a single sample, it demonstrates philosophical issues and possible misunderstandings in the use of maximum likelihood estimators and likelihood functions.

Given a discrete uniform distribution $1, 2, \ldots, N$ with unknown maximum, the UMVU estimator for the maximum is given by,

$$\frac{k+1}{k}m - 1 = m + \frac{m}{k} - 1$$

where m is the sample maximum and k is the sample size, sampling without replacement. This problem is commonly known as the German tank problem, due to application of maximum estimation to estimates of German tank production during World War II.

The formula may be understood intuitively as;

"The sample maximum plus the average gap between observations in the sample",

the gap being added to compensate for the negative bias of the sample maximum as an estimator for the population maximum.

This has a variance of $\dfrac{1}{k}\dfrac{(N-k)(N+1)}{(k+2)} \approx \dfrac{N^2}{k^2}$ for small samples $k \ll N$ so a standard deviation of

approximately N/k, the (population) average size of a gap between samples; compare $\dfrac{m}{k}$ above.

This can be seen as a very simple case of maximum spacing estimation.

The sample maximum is the maximum likelihood estimator for the population maximum, but, as discussed above, it is biased.

Applications

Numerous fields require the use of estimation theory. Some of these fields include (but are by no means limited to):

Measured data are likely to be subject to noise or uncertainty and it is through statistical probability that optimal solutions are sought to extract as much information from the data as possible.

Bayes' theorem is a formula that describes how to update the probabilities of hypotheses when given evidence. It follows simply from the axioms of conditional probability, but can be used to powerfully reason about a wide range of problems involving belief updates.

Given a hypothesis H and evidence E, Bayes' theorem states that the relationship between the probability of the hypothesis before getting the evidence $P(H)$ and the probability of the hypothesis after getting the evidence $P(H|E)$ is,

$$P(H|E) = \frac{P(E|H)}{P(E)} P(H).$$

Many modern machine learning techniques rely on Bayes' theorem. For instance, spam filters use Bayesian updating to determine whether an email is real or spam, given the words in the email. Additionally, many specific techniques in statistics, such as calculating $p$$p$-values or interpreting medical results, are best described in terms of how they contribute to updating hypotheses using Bayes' theorem.

Deriving Bayes' Theorem

Bayes' theorem centers on relating different conditional probabilities. A conditional probability is an expression of how probable one event is *given that* some other event occurred (a fixed value). For instance, "what is the probability that the sidewalk is wet?" will have a different answer than "what is the probability that the sidewalk is wet *given that* it rained earlier?" For a joint probability distribution over events A and B, $P(A \cap B)$, the conditional probability of A given B is defined as:

$$P(A|B) = \frac{P(A \cap B)}{P(B)}.$$

In the sidewalk example, where A is "the sidewalk is wet" and B is "it rained earlier," this expression reads as "the probability the sidewalk is wet given that it rained earlier is equal to the probability that the sidewalk is wet and it rains over the probability that it rains."

$P(A \cap B)$ is the probability of both A and B occurring, which is the same as the probability of A occurring times the probability that B occurs given that A occurred: $P(B|A) \times P(A)$. Using the same reasoning, $P(A \cap B)$ is *also* the probability that B occurs times the probability that A occurs given that BB occurs: $P(A|B) \times P(B)$. The fact that these two expressions are equal leads to Bayes' Theorem. Expressed mathematically, this is:

$$P(A\Big|B) = \frac{P(A \cap B)}{P(B)}, \text{if } P(B) \neq 0,$$

$$P(B\Big|A) = \frac{P(B \cap A)}{P(A)}, \text{if } P(A) \neq 0,$$

$$\Rightarrow P(A \cap B) = P(A\Big|B) \times P(B) = P(B\Big|A) \times P(A),$$

$$\Rightarrow P(A\Big|B) = \frac{P(B\Big|A) \times P(A)}{P(B)}, \text{if } P(B) \neq 0.$$

Notice that our result for dependent events and for Bayes' theorem are both valid when the events are independent. In these instances, $P(A\mid B) = P(A)$ and $P(B\mid A) = P(B)P(B\mid A) = P(B)$.

Bayes' Theorem:

$$P(A\mid B) = \frac{P(B\mid A)}{P(B)}P(A)$$

While this is an equation that applies to any probability distribution over events A and B, it has a particularly nice interpretation in the case where A represents a hypothesis H and B represents some observed evidence E. In this case, the formula can be written as:

$$P(H\mid E) = \frac{P(E\mid H)}{P(E)}P(H).$$

This relates the probability of the hypothesis before getting the evidence $P(H)P(H)$, to the probability of the hypothesis after getting the evidence, $P(H\mid E)$. For this reason, $P(H)$ is called the prior probability, while $P(H\mid E)$ is called the posterior probability. The factor that relates the two, $\frac{P(E\mid H)}{P(E)}$ is called the likelihood ratio. Using these terms, Bayes' theorem can be rephrased as "the posterior probability equals the prior probability times the likelihood ratio."

If a single card is drawn from a standard deck of playing cards, the probability that the card is a king is 4/52, since there are 4 kings in a standard deck of 52 cards. Rewording this, if King is the event "this card is a king the prior probability $P(King) = \frac{4}{52} = \frac{1}{13}$. If evidence is provided (for instance, someone looks at the card) that the single card is a face card, then the posterior probability $P(King\mid Face)$ can be calculated using Bayes' theorem:

$$P(King\mid Face) = \frac{P(Face\mid King)}{P(Face)}P(King).$$

Since every King is also a face card, $P(\text{Face}|\text{King}) = 1$. Since there are 3 face cards in each suit (Jack, Queen, King), the probability of a face card is $P(\text{Face}) = \dfrac{3}{13}$ using Bayes' theorem gives

$$P(\text{King}|\text{Face}) = \frac{13}{3}\frac{1}{13} = \frac{1}{3}.$$

Bayes' theorem clarifies the two-children problem from the first section:

- A couple has two children, the older of which is a boy. What is the probability that they have two boys?

- A couple has two children, one of which is a boy. What is the probability that they have two boys?

Define three events, A, B, and C, as follows:

  ABC = both children are boys
  B = the older child is a boy
  C = one of their children is a boy.

Question (A): is asking for $P(A|B)$ and Question (B): is asking for $P(A|C)$. The first is computed using the simpler version of Bayes' theorem:

$$P(A|B) = P(A)P(B|A) = \frac{\frac{1}{4}\cdot 1}{\frac{1}{2}} = \frac{1}{2}.$$

To find $P(A|C)$, we must determine $P(C)$, the prior probability that the couple has at least one boy. This is equal to $1 - P(\text{both children are girls}) = 1 - \dfrac{1}{4} = \dfrac{3}{4}$. Therefore the desired probability is

$$P(A|C) = \frac{P(A)P(C|A)}{P(C)} = \frac{\frac{1}{4}\cdot 1}{\frac{3}{4}} = \frac{1}{3}.$$

Visualizing Bayes' Theorem

Venn diagrams are particularly useful for visualizing Bayes' theorem, since both the diagrams and the theorem are about looking at the intersections of different spaces of events.

A disease is present in 5 out of 100 people, and a test that is 90% accurate (meaning that the test produces the correct result in 90% of cases) is administered to 100 people. If one person in the group tests positive, what is the probability that this one person has the disease?

The intuitive answer is that the one person is 90% likely to have the disease. But we can visualize this to show that it's not accurate. First, draw the total population and the 5 people who have the disease:

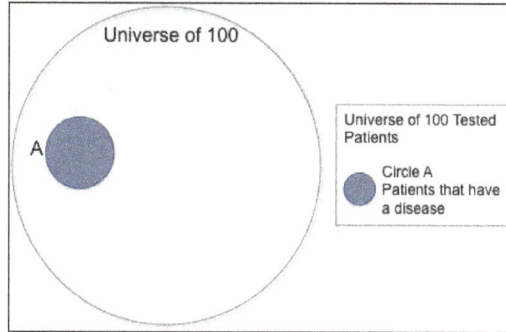

The circle A represents 5 out 100, or 5% of the larger universe of 100 people.

Next, overlay a circle to represent the people who get a positive result on the test. We know that 90% of those with the disease will get a positive result, so need to cover 90% of circle A, but we also know that 10% of the population who does not have the disease will get a positive result, so we need to cover 10% of the non-disease carrying population (the total universe of 100 less circle A).

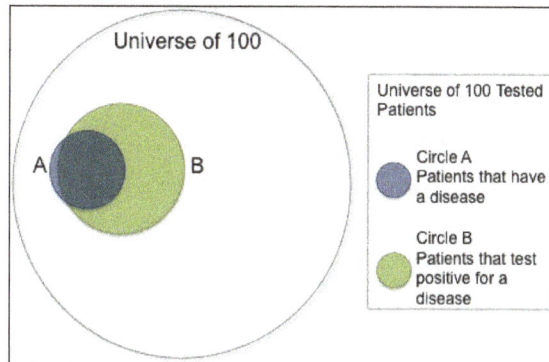

Circle B is covering a substantial portion of the total population. It actually covers more area than the total portion of the population with the disease. This is because 14 out of the total population of 100 (90% of the 5 people with the disease + 10% of the 95 people without the disease) will receive a positive result. Even though this is a test with 90% accuracy, this visualization shows that any one patient who tests positive (Circle B) for the disease only has a 32.14% (4.5 in 14) chance of actually having the disease.

Probability Axioms

The Kolmogorov axioms are a fundamental part of Andrey Kolmogorov's probability theory. In it, the probability P of some event E, denoted $P(E)$, is usually defined as to satisfy these axioms. The axioms are described below.

These assumptions can be summarised as follows: Let (Ω, F, P) be a measure space with $P(\Omega) = 1$. Then (Ω, F, P) is a probability space, with sample space Ω, event space F and probability measure P.

An alternative approach to formalising probability, favoured by some Bayesians, is given by Cox's theorem.

Axioms

First Axiom

The probability of an event is a non-negative real number:

$$P(E) \in \mathbb{R}, P(E) \geq 0; \qquad \forall E \in F$$

where F is the event space. It follows that $P(E)$ is always finite, in contrast with more general measure theory. Theories which assign negative probability relax the first axiom.

Second Axiom

This is the assumption of unit measure: that the probability that at least one of the elementary events in the entire sample space will occur is 1.

$$P(\Omega) = 1$$

Third Axiom

This is the assumption of σ-additivity:

Any countable sequence of disjoint sets (synonymous with *mutually exclusive* events) $E_1, E_2, \ldots$ satisfies:

$$P\left(\bigcup_{i=1}^{\infty} E_i \right) = \sum_{i=1}^{\infty} P(E_i).$$

Some authors consider merely finitely additive probability spaces, in which case one just needs an algebra of sets, rather than a σ-algebra. Quasiprobability distributions in general relax the third axiom.

From the Kolmogorov axioms, one can deduce other useful rules for calculating probabilities.

The Probability of the Empty Set

$$P(\varnothing) = 0$$

In some cases, $\varnothing$ is not the only event with probability 0.

Monotonicity

$$\text{if} \quad A \subseteq B \quad \text{then} \quad P(A) \leq P(B).$$

If A is a subset of, or equal to B, then the probability of A is less than, or equal to the probability of B.

The Numeric Bound

It immediately follows from the monotonicity property that:

$$0 \le P(E) \le 1 \qquad \forall E \in F.$$

Proofs

The proofs of these properties are both interesting and insightful. They illustrate the power of the third axiom, and its interaction with the remaining two axioms. When studying axiomatic probability theory, many deep consequences follow from merely these three axioms. In order to verify the monotonicity property, we set $E_1 = A$ and $E_2 = B \setminus A$, where $A \subseteq B$ and $E_i = \varnothing$ for $i \ge 3$. It is easy to see that the sets E_i are pairwise disjoint and $E_1 \cup E_2 \cup \cdots = B$. Hence, we obtain from the third axiom that:

$$P(A) + P(B \setminus A) + \sum_{i=3}^{\infty} P(E_i) = P(B).$$

Since the left-hand side of this equation is a series of non-negative numbers, and since it converges to $P(B)$ which is finite, we obtain both $P(A) \le P(B)$ and $P(\varnothing) = 0$. The second part of the statement is seen by contradiction: if $P(\varnothing) = a$ then the left hand side is not less than infinity:

$$\sum_{i=3}^{\infty} P(E_i) = \sum_{i=3}^{\infty} P(\varnothing) = \sum_{i=3}^{\infty} a = \begin{cases} 0 & \text{if } a = 0, \\ \infty & \text{if } a > 0. \end{cases}$$

If $a > 0$ then we obtain a contradiction, because the sum does not exceed $P(B)$ which is finite. Thus, $a = 0$. We have shown as a byproduct of the proof of monotonicity that $P(\varnothing) = 0$.

Another important property is:

$$P(A \cup B) = P(A) + P(B) - P(A \cap B)$$

This is called the addition law of probability, or the sum rule. That is, the probability that A or B will happen is the sum of the probabilities that A will happen and that B will happen, minus the probability that both A and B will happen. The proof of this is as follows:

Firstly,

$$P(A \cup B) = P(A) + P(B \setminus A)$$

So,

$$P(A \cup B) = P(A) + P(B \setminus (A \cap B)) \text{ (by } B \setminus A = B \setminus (A \cap B) \text{)}.$$

Also,

$$P(B) = P(B \setminus (A \cap B)) + P(A \cap B)$$

and eliminating $P(B \setminus (A \cap B))$ from both equations gives us the desired result.

An extension of the addition law to any number of sets is the inclusion–exclusion principle.

Setting B to the complement A^c of A in the addition law gives:

$$P\left(A^c\right) = P(\Omega \setminus A) = 1 - P(A)$$

That is, the probability that any event will *not* happen (or the event's complement) is 1 minus the probability that it will.

Simple Example: Coin toss

Consider a single coin-toss, and assume that the coin will either land heads (H) or tails (T) (but not both). No assumption is made as to whether the coin is fair.

We may define:

$$\Omega = \{H, T\}$$

$$F = \{\varnothing, \{H\}, \{T\}, \{H, T\}\}$$

Kolmogorov's axioms imply that:

$$P(\varnothing) = 0$$

The probability of *neither* heads *nor* tails, is 0.

$$P(\{H, T\}^c) = 0$$

The probability of *either* heads *or* tails, is 1.

$$P(\{H\}) + P(\{T\}) = 1$$

The sum of the probability of heads and the probability of tails, is 1.

References

- Probability, science: britannica.com, Retrieved 25 February, 2019

- Roe RM, Busemeyer JR, Townsend JT (2001). "Multialternative decision field theory: A dynamic connectionst model of decision making". Psychological Review. 108 (2): 370–392. doi:10.1037/0033-295X.108.2.370

- Markovian-processes, probability-theory, science: britannica.com, Retrieved 26 March, 2019

- V.G.Voinov, M.S.Nikulin, "Unbiased estimators and their applications. Vol.2: Multivariate case", Kluwer Academic Publishers, 1996, ISBN 0-7923-3939-8

- Bayes-theorem: brilliant.org, Retrieved 27 April, 2019

PERMISSIONS

INDEX

A

Abstract Algebra, 12, 34, 58, 65, 92, 112, 114-115
Acute Angle, 154
Algebraic Equation, 76, 113, 122
Algebraic Expression, 93, 127
Algebraic Geometry, 65, 105, 139
Analytic Geometry, 3, 28, 64, 72-74, 76-77, 92, 108, 121, 129
Analytic Trigonometry, 121, 123, 129
Austrian Method, 38-39

B

Binomial Distribution, 182, 188, 192, 195-196, 198, 200, 206-207, 213
Boolean Algebra, 115

C

Calculus, 3, 10, 12, 29-30, 45, 58, 71, 83-84, 92, 94, 100, 110, 129, 155-156, 159, 163, 176, 182, 207, 210
Cartesian Coordinates, 73, 76, 130-131, 156
Cartesian Plane, 74, 138
Category Theory, 11, 106, 112
Circular Arc, 124
Complex Numbers, 42, 48-49, 51, 57-58, 65, 94, 98-104, 107-109, 122, 142
Conditional Probability, 190-192, 198, 211-212, 228
Congruence Of Triangles, 67-68
Congruent Number, 136
Continuous Functions, 10, 163, 232
Corresponding Angles, 68, 128, 158

D

Decimal Fraction, 55-56, 61-62
Decimal Notation, 59
Decision Theory, 15, 172, 219-222
Derivative, 3, 58, 80, 84, 99, 145, 209-210, 226
Differential Calculus, 10, 83
Differential Equation, 160, 209
Diophantine Equations, 27, 112
Distributive Law, 20, 109
Dot Product, 43, 75, 77

E

Euclidean Algorithm, 22, 26, 56, 141, 143
Euclidean Division, 52, 55-56
Euclidean Geometry, 8, 64-66, 70-71, 76, 80, 97
Euler Number, 161

F

Fibonacci Number, 149

G

Galois Theory, 95-97, 104, 158
Gaussian Curvature, 86
Gaussian Integers, 98, 142-144

H

Hyperbolic Geometry, 80-83

I

Infinite Product, 122
Inverse Trigonometric Functions, 117, 166, 169-170
Irrational Numbers, 24, 33, 108, 123, 127

L

Law Of Sines, 121, 127-129, 166-167
Line Segment, 24, 35, 66, 68, 88, 93, 124-125, 127, 187
Linear Equations, 2, 74, 100-101, 112

M

Matrix Theory, 100-101
Multiplicative Inverse, 49, 55

N

Natural Numbers, 18-20, 35, 48-50, 52-53, 98, 103, 136, 158, 204
Number Theory, 2-3, 8-9, 12, 16, 18, 24-33, 96-98, 101, 113, 115

P

Parametric Equation, 138
Pell Equation, 27, 148
Periodic Functions, 153, 157, 164, 168
Plane Geometry, 2, 67, 71

Plane Trigonometry, 117-118, 127, 129

Polar Coordinates, 73, 130

Polar Equation, 130

Polynomial Equations, 65, 93, 102-103, 112

Polynomial Ring, 51

Positional Numeral System, 58, 60

Prime Factorization, 98-99

Prime Number Theorem, 30-31

Probability, 3, 15-16, 28, 172-178, 180-182, 184-208, 211-218, 220-223, 225, 228-234

Projective Geometry, 3, 64, 87, 89, 97

Pythagorean Theorem, 24, 66, 69-70, 76-77, 117-119, 132, 143, 152-153, 164, 167

Pythagorean Triple, 117, 132-138, 141-146, 148-149, 151

Q

Quadratic Equation, 75

R

Rational Numbers, 23-24, 42, 50-53, 56, 60, 62, 99, 103, 106, 138-139, 201

Real Numbers, 3, 27, 34-36, 42, 50-52, 55, 57-58, 61, 65, 73, 75, 102-103, 107-108, 122, 156, 187, 194, 200, 202-203

Regular Polygons, 69-71

Repeating Decimal, 59, 62

Riemannian Geometry, 65, 82

S

Scatter Plot, 132-133, 144-145

Set Theory, 50, 115

Solid Geometry, 2, 66, 71

Spherical Trigonometry, 117, 121, 128-129

Structural Approach, 104-106

T

Tangent Line, 80, 156

Trigonometry, 1-2, 9, 68, 117-124, 127-129, 139, 168, 170-171

U

Unit Circle, 78, 124, 138-139, 153-158

V

Vector Multiplication, 43

Vector Spaces, 69, 112

W

Whole Numbers, 6, 18, 20, 22, 26-29, 31, 37, 42, 51

Z

Zero Factor Property, 110